AMERICAN PHILOSOPHICAL SOCIETY

HELD AT PHILADELPHIA

FOR

PROMOTING USEFUL KNOWLEDGE

CELEBRATION

OF THE

HUNDREDTH ANNIVERSARY

MAY 25, 1843

New Foreword by Robert M. Hauser

275 YEARS AT THE
AMERICAN PHILOSOPHICAL SOCIETY

PROCEEDINGS

OF THE

AMERICAN PHILOSOPHICAL SOCIETY

HELD AT PHILADELPHIA

FOR

PROMOTING USEFUL KNOWLEDGE

VOL. III

No. 27. MAY 25–30, 1843

PHILADELPHIA
PRINTED FOR THE SOCIETY
By John C. Clark, 60 Dock Street
1843

Transactions of the
American Philosophical Society
Held at Philadelphia
For Promoting Useful Knowledge
Volume 107, Part 5

ISBN: 978-1-60618-075-4

U.S. ISSN: 0065-9746

Library of Congress Cataloging-in-Publication Data

Names: American Philosophical Society.
Title: Celebration of the hundredth anniversary of the American
    Philosophical Society : proceedings of the American Philosophical
    Society, 1843.
Other titles: Celebration of the 100th anniversary of the American
    Philosophical Society
Description: Philadelphia : American Philosophical Society Press,
    [2019] | Series: Transactions of the American Philosophical
    Society, ISSN 0065-9746; volume 107, part 5 | Includes index.
Identifiers: LCCN 2019019781 | ISBN 9781606180754 (alk. paper)
Subjects: LCSH: American Philosophical Society—Anniversaries, etc.
    | Science—United States—History—19th century.
Classification: LCC Q11.A44 C45 2019 | DDC 509.73—dc23
LC record available at https://lccn.loc.gov/2019019781

CONTENTS

For historical accuracy, the material presented on pages 164–169 is exactly as published by the American Philosophical Society in the *Proceedings* in 1843. This text is patently racist and discriminatory and in no way represents the present beliefs or practices of the American Philosophical Society.

NOTE. By a resolution of the Society, the charge and entire responsibility of the published" "Proceedings" are devolved on one of the Secretaries, who acts as Reporter. That officer has been aided, however, in the present number by several gentlemen, who have furnished him with abstracts of their contributions to the meeting; and he mentions the fact, at once to tender to them his acknowledgments for their courtesy, and to relieve himself from the censure of having compressed some of the most interesting communications within too narrow limits.

Hall of the American Philosophical }
Society, September 10, 1843. }

FOREWORD

In 1843, the American Philosophical Society celebrated the 100th anniversary of its founding by Benjamin Franklin. It is fitting, then, that in celebrating the 275th year of the Society, it reissues the celebratory volume of the APS *Proceedings* that accompanied the centennial. That year marked more than one turning point in the history of the Society, for it was, also, the last year of the life of its long-time President, the remarkable lawyer and linguist, Peter S. Du Ponceau. At the age of 17, Du Ponceau had emigrated from France to New Hampshire in 1777 as an aide to Baron Friedrich Wilhelm Von Steuben, the Prussian general who was a key leader of the Continental Army during the American Revolutionary War. Thus, Du Ponceau's death marked the passing of one of the last survivors of the Revolutionary era as well as an early Member of the Society.

This volume lists 322 Members of the Society, of whom 94—well over 25 percent—were international members. Among those, the French (22) and English (19) were the most numerous, whereas many other nations were represented in smaller numbers: Argentina, Austria, Belgium, Canada, Denmark, Guatemala, Hungary, Ireland, Italy, Macao, Mexico, the Netherlands, Portugal, Prussia, Russia, Scotland, Spain, and Sweden. Of 228 domestic members, a substantial majority, 134, were Pennsylvanians. Massachusetts (21), New York (15), and New Jersey (8) were also well represented. The dominance of eastern states among the Membership stands in sharp contrast to the present, bicoastal distribution of the Members of the Society.

Benjamin Chew, Jr., was earliest elected among the Society's Members of 1843. He was a prominent lawyer, statesman, land speculator, and friend of George Washington. Chew's election in 1787 predated that of Du Ponceau by four years. Many influential and familiar names (to this writer) occur in the list of Members (listed by date of election):

- Albert Gallatin (1791)
- Peter Stephen Du Ponceau (1791)
- Alexander von Humboldt (1804)
- John Quincy Adams (1818)
- James Fenimore Cooper (1823)
- Joel R. Poinsett (1827)
- Noah Webster (1827)
- Samuel George Morton (1828)
- Alexander Dallas Bache (1829)
- Washington Irving (1829)
- John James Audubon (1831)
- Titian R. Peale (1833)
- Franklin Peale (1833)
- Joseph Henry (1835)
- Thomas Sully (1835)
- Daniel Webster (1837)
- Adolphe Quetelet (1839)
- Thomas U. Walter (1839)
- Michael Faraday (1840)
- Alexis de Tocqueville (1842)

No doubt, other names will be familiar to other readers.

Contributions to the centennial *Proceedings* were, with only a couple of exceptions, every bit as varied as those to today's semi-annual meetings of the Society. Themes of the many contributions included astronomy, meteorology, geology, fossils, mathematics, chemistry, finance, scientific instrumentation and measurement, biography, Colonial history, medicine, mortality, and geographic exploration. The obvious exceptions are the absence of contemporary contributions in the arts and humanities and discussions of contemporary social and political issues.

Many of the contributions are long-since scientifically obsolete. For example, despite the interest of Members in earlier forms of life, the Society's centennial preceded Darwin's *Origin of Species* by more than a dozen years. In concluding remarks on the history of the Society, Isaac Hays—a leading scientist of the day and active APS Member—epitomized a pre-Darwinian understanding of the world: "Dr. Hays concluded with some remarks on the changes that have taken place in organized bodies, corresponding to the change in the condition of the surface of the Earth—on the proofs they afford not only of an intelligence, adapting mechanism to an end, but of successive manifestations of the same contriving intelligence, adjusting the mechanism to the altered conditions under which it was to exist—and on the reverential and exalted ideas this ought to impress on us respecting the wisdom, power, and goodness of the Creator, by whose fiat all things are called into existence and made to perish, and who alone endures forever, and whose years have no end" (p. 53).

Two contributions are egregious for their patently racist content: Samuel George Morton's "Summary of his Series of Observations on Egyptian Ethnography" (pp. 129–33) and Benjamin Coate's memoir "On the Effects of Secluded and Gloomy Imprisonment on Individuals of the African Variety of Mankind in the Production of Disease" (pp. 164–69). Of course, this writer and, I expect, all present Members of the Society find the racist sentiments expressed in these two texts to be both false and abhorrent. However, although we briefly considered the possibility of deleting these two passages from the reissued *Proceedings*, our decision was to present the full text of this mid-19th century compendium with an appropriate disclaimer and warning to present-day readers. Thus, the reissued *Proceedings* of 1843 reflects both the strengths and weaknesses of American thought leaders of the mid-19th century.

Despite the corrective value of hindsight, the *Proceedings* of 1843 makes for a lively reading experience, from Isaac Hays's brief history of the Society all the way to Stephen Alexander's historical review and explanation of visual distortions in the observation of solar eclipses.

Robert M. Hauser
Executive Officer
October 2019

275 YEARS AT THE AMERICAN PHILOSOPHICAL SOCIETY

American Philosophical Society

CELEBRATION

OF THE

HUNDREDTH ANNIVERSARY,

MAY 25, 1843.

This being the hundredth anniversary of the organization of the American Philosophical Society under that title, the members, and a large number of guests invited by them, assembled at noon in the saloon of the Musical Fund Society, for the purpose of celebrating it.

The exercises of the day were opened by the President, Mr. Du Ponceau, in the following

ADDRESS.

We are assembled here to celebrate the hundredth anniversary of the formation of our Association, under the name of the American Philosophical Society, which it still bears, and which I hope it will long continue to preserve. In the regular course of things, it would have been my duty to address you on this occasion, and nothing could have afforded me greater pleasure; but, gentlemen, in the evening of life, the power does not always follow the will.

You have, however, no reason to regret my deficiency; for the important duty is about to be performed by a gentleman, whose talents and capacity far exceed the ability for the task which I might at any time have claimed. He will relate to you the interesting history of our Association from its beginning, and the slow, but undeviating steps, by which it has risen to the distinguished rank which it now occupies among the learned societies of the civilized world. But before we proceed farther, it is necessary that we should recollect that, "every good gift is from above, and cometh down from the Father of lights, with whom is no variableness, neither shadow of turning." It is right and meet, therefore, that we should begin by invoking the blessing of the Almighty Creator, that he may impart to us those perfect gifts which he alone can bestow, and direct our future labors to his immortal glory, and to the happiness of mankind. For this reason, I beg leave to call on the Rev. Provost of the University of Pennsylvania, here present, to perform that sacred duty, and to turn our minds to the great Source of knowledge, before we begin to speak of our humble efforts to promote its advancement.

The Reverend Dr. Ludlow, Provost of the University of Pennsylvania, then offered the following

PRAYER.

Almighty and most merciful God, who art the fountain of all being and blessing, the creator, preserver, and governor of all things, we revere thy glorious Majesty, and, with penitent and grateful hearts, acknowledge thy bountiful goodness to us, and to all men. We beseech thee, whose inspiration hath given us understanding, so to enlighten our minds in the study of thy works of nature and of grace, that we may have new and clearer discoveries of thy glorious perfections, and, above all, that we may know thee, through Jesus Christ, unto eternal life. Regard, O Lord, with thy merciful favor, the cause of science, as well as of religion. May both be constantly and harmoniously advancing, that they may together tend to further thy wise purposes, and the best interests of the whole family of man. Especially vouchsafe thy continued goodness to this scientific institution, whose hundredth anniversary we this day celebrate, that its usefulness to our beloved country and the world may go on increasing, and that all its members, purified by the power of divine truth, may rise at last to behold and study thy works and ways where no cloud shall interpose to obscure thy glory. This mercy we ask, with the forgiveness of our sins, for the sake of our blessed Savior; and to God, only wise, be the praise, world without end. Amen.

The prayer being concluded, Dr. Patterson, one of the Vice-Presidents, pronounced the following

DISCOURSE.

A century is this day accomplished since the American Philosophical Society was first organized under that title, and we are met together to celebrate the event. When we reflect that it is not three hundred and fifty years since this continent was discovered, and that it is less than half that period since the spot where Philadelphia now stands was a forest occupied by savages, we see that our institution is one of no inconsiderable American antiquity. Our Society is, indeed, the oldest of its class in the new world; and it is not without a natural feeling of filial pride that we refer to the fact, that in the infancy of the country, and amid the cares and struggles of early colonization, there were found among our forefathers men who could employ themselves in the higher occupations of the mind, combine together for the advancement of knowledge, and feel and act upon the conviction, that science must always form one of the elements of honor and prosperity in a civilized state.

The occasion of our meeting has been deemed a suitable one for tracing the progress of our institution from its obscure and limited beginnings, through its struggles under different forms and different names, to the union which first gave it stability, and up to the time when, rising with the destinies of the country, it was able to take a station and a rank among the learned societies of the world. Such is the task that I am called upon to perform in the present address.

In this review, it is natural that we should attach a particular feeling of interest to the more remote periods; for societies, as well as individuals, are not without curiosity at least, and perhaps pride, in relation to their ancestry. Besides, the day we celebrate is of the olden time, and thus seems to call our special notice to events of early date. I shall not hesitate, therefore, to make the early history of our Society the prominent subject of this address.

Among the fathers of our city who possessed a love of knowledge and a desire to promote it, he that stood preeminent was the illustrious individual whom we claim as our founder—Benjamin Franklin. Possessing, as he did, skill, industry, and frugality, he might have safely walked alone through life, assured of personal success, and, like so many others, might have sought to dignify selfishness by calling it independence. But his course was far different. Without neglecting his own advancement, he still kept constantly in view that of the community in which he lived, and of his country, and he knew that such objects could only be attained by combined efforts. Hence the number of associations in this city of which he was either the founder, or the most efficient promoter. We have examples in the establishment of the first fire company, the first public library, the first hospital, and the first academy, now the University of Pennsylvania.

But of all the societies which he originated, the earliest was one which requires our particular notice on this occasion; for it is claimed, not without plausibility, as being itself one of the branches of which our present Society is composed, and it is at least the original after which that branch was modeled. I refer to the celebrated Junto, which was founded when Franklin was but 21 years of age. In the story of his life, told by himself so simply yet so attractively, and which is read in every part of the civilized world, he gives the following account of the origin of this association:

In the autumn of 1727, "I formed most of my ingenious acquaintance into a club for mutual improvement, which we called the *Junto.* We met on Friday evenings. The rules that I drew up required, that every member in his turn should produce one or more queries on any point of morals, politics, or natural philosophy, to be discussed by the company; and once in three months produce and read an essay of his own writing, on any subject he pleased. Our debates were to be under the direction of a president, and to be conducted in the sincere spirit of inquiry after truth, without fondness for dispute, or desire of victory; and to prevent warmth, all expressions of positiveness in opinions, or direct contradiction, were after some time made contraband, and prohibited under small pecuniary penalties."

Franklin then gives the names of the first members, with brief outlines of their characters, drawn in his peculiar and happy style. Two of the individuals named in this list were afterwards among the original members of the Philosophical Society. These are William Coleman and Thomas Godfrey.

The first is described by Franklin as having "the coolest, clearest head, the best heart, and the exactest morals of almost any man he had ever met with. He afterwards became a merchant of great note, and one of the provincial judges."

The second is thus introduced: "Thomas Godfrey, a self-taught mathematician, and afterwards inventor of what is now called *Hadley's Quadrant.* But he knew little out of his way, and was not a pleasing companion; as like most great mathematicians I have met with, he expected universal precision in every thing said, or was forever denying and distinguishing upon trifles, to the disturbance of all conversation."

The claim of Godfrey over Hadley to the invention of that valuable instrument the quadrant, seems to be fully established by our fellow-member, Dr. Miller, in his Retrospect of the Eighteenth Century. But a third party to the claim has been brought forward in the person of Sir Isaac Newton, and he has seized upon the lion's share. It appears that, after the death of Halley, a description of such an instrument was found among his papers, by his executor, in Newton's own handwriting, and was communicated to the Royal Society twenty-five years after Newton's death, which occurred in 1727. It is certain, therefore, that Godfrey cannot have had any knowledge of this paper, and it is matter of regret that it should interfere, in any degree, with his claim to originality in an invention which would make a reputation for any ordinary man, while it can scarcely add to that of Newton.

I hope I may be permitted to protest against any general conclusion being drawn from Franklin's experience of the captiousness of mathematicians; and to the example of one of the earliest of our American geometers, I would oppose that of one of the last and most distinguished—Nathaniel Bowditch—who was a man remarkable for his social virtues, modest and attractive manners, and Franklinian common sense.

The Junto was, properly speaking, a debating society. At first it met at a tavern; but subsequently at the house of one of the members, Robert Grace, whom Franklin characterizes as "a gentleman of some fortune, generous, lively, and witty, a lover of punning and of his friends." I am happy to say that Robert Grace is not without his successors in our present Society.

One of the rules of the Club was that the institution should be kept a secret; the intention being, as Franklin states, to avoid applications of improper persons for admittance. The number of members at any one time was limited to twelve, but vacancies were filled as they occurred, and the names of twenty-three members are preserved.

On admission into the Club, a course was followed which is too remarkable in itself, and in its bearing upon a difficult question in the history of this Society, not to be here introduced. It is thus presented in Franklin's papers:

"Any person to be qualified—to stand up, and lay his hand upon his breast, and be asked these questions, viz:

"1st. Have you any particular disrespect to any present member? Answer: *I have not.*

"2d. Do you sincerely declare that you love mankind in general, of what profession or religion soever ? Ans. *I do.*

"3d. Do you think any person ought to be harmed in his body, name, or goods, for mere speculative opinions, or his external way of worship'? Ans. *No.*

"4th. Do you love truth for truth's sake, and will you endeavor impartially to find and receive it yourself, and communicate it to others? Ans. *Yes.*"

No minutes of the proceedings of the original Junto are preserved, but Franklin mentions in his autobiography several questions of great interest which were discussed at it, and several pieces read before it and afterwards published in his newspaper.

It was at one time proposed to increase the number of members; but to this Franklin was opposed, and instead of it he made "a proposal that every member separately should form a subordinate club, with the same rules respecting queries, &c., and without informing them of the connexion with the Junto." "This project was approved, and every member undertook to form a club; but they did not all succeed. Five or six only were completed, which were called by different names, as, the *Vine*, the *Union*, the *Band*." Of these subordinate companies, a brief paragraph in Franklin's Life is the only remaining record.

In speaking of William Coleman, Franklin says: "Our friendship continued without interruption to his death, upwards of forty years; and the Club continued almost as long, and was the best school of philosophy, morality, and politics, that then existed in the province."

While Franklin was abroad, he shows by his correspondence that he still held the institution of his youth in affectionate remembrance. This appears repeatedly in his letters to his friend Hugh Roberts. He calls it "the good old Club," "the ancient Junto." So late as 1765, he says: "I wish you would continue to meet the Junto, notwithstanding that some effects of our political misunderstanding may sometimes appear there. It is now perhaps one of the oldest clubs, as I think it was formerly one of the best, in the king's dominions." Even in 1766, he writes:—"Remember me affectionately to the Junto."

It appears, then, that the Junto continued in existence about forty years. But did it keep up its original character? This may well be doubted. The members grew gradually to be old men, and it is hardly to be supposed that they would submit to the task of writing essays, or would formally propose questions, and afterwards debate them. Their fortunes were made, their education completed; and it is therefore much more probable, that when the remnant of the once youthful and active Junto met together, they indulged themselves in social conversation and temperate conviviality. Such is said to be the tradition in the Roberts family; and it is confirmed by a letter from Dr. Franklin to their ancestor, written in 1761, in which he says:—"You tell me you sometimes visit the ancient Junto. I wish you would do it oftener. Since we have held that Club till we are grown gray together, let us hold it out to the end. For my own part, I find I love company, chat, a laugh, a glass, and even a song, as well as ever; and at the same time relish better than I used to do the grave observations and wise sentences of old men's conversation; so that I am sure the Junto will be still as agreeable to me as it ever has been. I therefore hope it will not be discontinued, as long as we are able to crawl together."

In May 1765, Hugh Roberts writes as follows to Dr. Franklin:—"I sometimes visit the worthy remains of the ancient Junto, for whom I have a high esteem; but alas, the political, polemical divisions have in some measure contributed to lessen that harmony we there formerly enjoyed." To this letter, Franklin answers in July following, urging his friend's attendance at the Junto, almost in the same terms used some years before, and which we have just quoted, and then closes his exhortation in the following touching words:—"We loved and still love one another. We are grown gray together, and yet it is too early to part. Let us sit till the evening of life is spent. The last hours are always the most joyous. When we can stay no longer, it is time enough then to bid each other good night, separate, and go quietly to bed."

Such appears to be the true history of this memorable Club, at least so far as regards its early members. In its youth it was full of activity and ambition: in its age, "frosty but kindly," these were blunted, but the social affections remained, and the old friends were happy to meet together, talk over the past, and enjoy the present. Examples of the same kind have not been uncommon; and Franklin's own Fire Company, the *Union*, was one. It was continued as a social club, long after age had unfitted the members for the active duties of firemen. Indeed, I understand that its organization still continues, although it is very long since it has possessed an engine, or has pretended to aid at fires.

As there was a provision for filling up the places in the Junto as they fell vacant, the means existed of continuing the Club in constant vigor; and it may naturally be asked, whether this succession was not kept up, and for what time, and whether with younger members, the original rules of proceeding may not have been enforced. We shall have occasion to consider this question in another part of our narrative; but we have now to speak of the origin of an institution of a different character.

The Junto was, as we have seen, a secret society, limited to a small number of members, local in its character, and confined in its objects. It did not claim to be a scientific body. But the British provinces were fast advancing in population and prosperity, and the active mind of Franklin, ever seeking to do good, and especially anxious for the promotion of knowledge, became impressed with the importance of establishing a national institution for the cultivation of science; and he accordingly issued and distributed a proposal for this purpose, in the form of a printed circular. This proposal was the true origin of the *American Philosophical Society*. It bears the date of the 14th of May 1743, old style, corresponding in the Gregorian calendar to the 25th, on which we are now met, after the lapse of one hundred years, to celebrate the birthday of our Institution.

Dr. Franklin's circular is entitled "A proposal for promoting useful knowledge among the British Plantations in America." It commences by speaking of the great extent of the Colonial possessions, "having different climates and different soils, producing different plants, mines, and minerals, and capable of different improvements, manufactures," &c.

It then says: "The first drudgery of settling new colonies, which confines the attention of people to mere necessaries, is now pretty well over; and there are many in every province in circumstances that set them at ease, and afford leisure to cultivate the finer arts, and improve the common stock of knowledge. To such of these who are men of speculation, many hints must from time to time arise, many observations occur, which if well examined, pursued, and improved, might produce discoveries to the advantage of some or all of the British Plantations, or to the benefit of mankind in general. . . . But as, from the extent of the country, such persons are widely separated, and seldom can see and converse or be acquainted with each other, so that many useful particulars remain uncommunicated, die with the discoverers, and are lost to mankind; it is to remedy this inconvenience for the future, proposed,—

"That one society be formed of *virtuosi*, or ingenious men, residing in the several colonies, to be called *The American Philosophical Society*, who are to maintain constant correspondence.

"That Philadelphia, being the city nearest to the centre of the continent colonies, communicating with all of them northward and southward by post, and with all the islands by sea, and having the advantage of a good growing library, be the centre of the Society.

"That at Philadelphia there be always at least seven members, viz. a physician, a botanist, a mathematician, a chemist, a mechanician, a geographer, and a general natural philosopher, besides a president, treasurer, and secretary.

"That these members meet once a month, or oftener, at their own expense, to communicate to each other their observations and experiments; to receive, read, and consider such letters, communications, or queries as shall be sent from distant members; to direct the dispersing of the copies of such communications as are valuable, to other distant members, in order to procure their sentiments thereupon."

Then follows an enumeration, made with some detail, of the subjects on which it was proposed that the Society should be occupied: including investigations in botany; in medicine; in mineralogy and mining; in mathematics; in chemistry; in mechanics; in arts, trades, and manufactures; in geography and topography; in agriculture; and "all philosophical experiments that let light into the nature of things, tend to increase the power of man over matter, and multiply the conveniences or pleasures of life."

The circular proposes that "a correspondence be kept up with the Royal Society of London, and the Dublin Society; that abstracts of the communications be sent quarterly to all the members; and "that, at the end of every year, collections be made and printed of such experiments, discoveries, and improvements, as may be thought of public advantage."

The duties of the secretary are particularly laid down, and they are very arduous; requiring that he attend to all the correspondence, "abstract, correct, and methodize such papers as require it, and as he shall be directed to do by the president, after they have been considered, debated, and digested in the Society; to enter copies thereof in the Society's books, and make out copies for distant members." And after enumerating these difficult duties, the circular closes by saying:

"Benjamin Franklin, the writer of this proposal, offers himself to serve the Society as their Secretary, till they shall be provided with one more capable."

In this projêt we find all the leading features of our present Society. There can be no doubt that from the day when it was proposed the necessary measures for carrying it into execution were taken. Dr. Thomas Bond (himself one of the original members,) in an oration delivered before our Society in 1782, says:—"Franklin gradually established many necessary institutions, among which was this Philosophical Society, so early as 1743, when the plan was formed and published, the members chosen, and an invitation given to all ingenious persons to cooperate and correspond with them on the laudable occasion." It is true that Franklin, in his autobiography, gives the date 1744, saying, "in that year I succeeded in proposing and establishing a Philosophical Society. The paper I wrote for that purpose will be found among my writings, if not lost with many others." But Franklin wrote from memory, and the date of the paper referred to, which was doubtless the proposal of 1743, shows that he had made a mistake in the year.

In a letter to Cadwallader Colden, dated New York, 5th April, 1744, Dr. Franklin acquaints him "that the Society, as far as relates to Philadelphia, was actually formed, and had had several meetings to mutual satisfaction."

In this letter the following list is presented of the original members:

DR. THOMAS BOND, as Physician.
MR. JOHN BARTRAM, as Botanist.
MR. THOMAS GODFREY, as Mathematician.
MR. SAMUEL RHOADS, as Mechanician.
MR. WILLIAM PARSONS, as Geographer.
DR. PHINEAS BOND, as General Natural Philosopher.
MR. THOMAS HOPKINSON, President.
MR. WILLIAM COLEMAN, Treasurer.
BENJAMIN FRANKLIN, Secretary.

We have here many distinguished names.

The first president, Thomas Hopkinson, is said to have "possessed a fine genius and a finished education, having been a student at Oxford. He was born in London in 1709, and came while young to Philadelphia, where he died at the early age of forty-two." Being fond of the pursuits of science, as well as of letters, he often assisted in the electrical and philosophical experiments of Franklin, who, in one of his letters introduces the following note respecting him: —"The power of points to *throw off* the electrical fire was first communicated to me by my ingenious friend Mr. Thomas Hopkinson, since deceased, whose virtue and integrity in every station of his life, public and private, will ever make his memory dear to those who knew him, and knew how to value him." Doctors Thomas and Phineas Bond were eminent and learned men. The former of them was the original projector of the Pennsylvania Hospital, though he failed in his efforts to establish it, until he had recourse to the sagacity and powerful influence of Franklin. John Bartram was the founder of the Botanic Garden in our vicinity, which still exists, and bears his name. Thomas Godfrey is spoken of by Franklin as a great mathematician, always absorbed in his studies. Of his invention of the quadrant, I have already spoken.

There can be no doubt that the plan of establishing the Philosophical Society had been often brought before the Junto for consideration, for we know that it was the practice of Franklin, when he had new projects to propose, to have them first discussed in the Club. But a stronger evidence still of the part which they took in forming the new institution is presented by the fact that of the nine original members of the Philosophical Society, six, including the three officers, are known to have belonged to the Junto,—namely, Franklin, Hopkinson, Coleman, Godfrey, Rhoads, and Parsons.

The minutes of the early proceedings of the Philosophical Society are not preserved, and we are left in the dark, not only as to its labors, but as to the time that it remained in activity. It is only from indirect evidence that we are able to infer that it did not continue its meetings for more than ten years, when it went into a state of suspended animation, from which it was destined to revive, at a future period, and to flourish with greater vigor than it had ever possessed in the earliest days of its existence.

But before we speak of this revival, our attention must be directed to another institution, which came, in the end, to take its equal share in the formation of our present Society. It was at first, and for many years, called the *Junto*; and this name at once draws out attention, and leads us to inquire whether it was identical with the ancient Junto, or if not, what was the relation between them.

We are, fortunately, in possession of the minutes of the new Association from 1758 to 1768 inclusive, though with some intermissions; and from these and the Franklin correspondence, the following points may be established. And, first, I shall present the circumstances, which tend to show an identity between the two Juntos.

Of these, the most striking is the name itself. Our venerable President, who laid, before the Society, three years ago, a most interesting and able paper on its early history, took the ground that there can, in fact, have been but one Junto, and remarks, with great force, that "it is hardly credible that while the old Junto existed, another society should have adopted the same name," as "it would have been contrary to all the rules of delicacy and mutual respect."

Another point of correspondence is, that the number of members of the two Juntos was the same, namely, twelve.

A third is that they both met on the evenings of Friday, a time that has been so set apart by the different societies, founded by Franklin for promoting useful knowledge, from 1724 to the present day.

Fourthly, the two Clubs (for so they both call themselves) had the same objects, and pursued them in the same manner. In each, the members sought their "mutual improvement," by reading essays, and proposing and discussing questions.

Fifthly, they were both secret societies.

Lastly, a most striking coincidence is presented in the fact, that the four qualifications for admission, required of new members on their initiation, were the same in both Juntos.

Were these the only facts known respecting the new Junto, we should not hesitate to conclude that it was but a continuation of the old, which had been kept up, as in other societies, by the successive admission of new members, as the older ones died or resigned. But there are other circumstances which are inconsistent with this view of the case, and which it is therefore necessary to consider.

One of these is the evidence of a distinct date for the origin of the new Junto. In a letter to Dr. Franklin, filled with interesting matter respecting this institution, written on the 6th of November, 1768, by Charles Thomson, (afterwards so well known as Secretary of the Congress of the Revolution) he introduces his subject by saying: "You remember the Society to which I belonged, which was begun in the year 1750." This, then, fixes the year. But again: In the minutes of the Junto, several notices are given of the celebration of its anniversary, when (in the words of the record) "it was customary for them to express their good wishes for the Society's prosperity," at an entertainment provided for the occasion, and when "the original laws were distinctly read, and also the by-laws." Now this anniversary was, at least for many years, held on the 5th of February. It appears, then, that the institution must have had its origin certainly in 1750, and probably on the 5th of February of that year. But we learn, from the evidence of Franklin, that the old Junto was formed in the autumn of 1727. The two Juntos, therefore, cannot be the same.

Another proof of the want of identity of the two Juntos is found in a comparison of the lists of members. From the year 1758 the roll of the new Junto is recorded in their minutes, where the absent members are named as well as the present. Now it is a significant fact, that in these lists there is not found the name of a single member of the old Club, although five of them at least were living in 1758, viz., Franklin, Coleman, Hugh Roberts, Philip Syng, sen., and Samuel Rhoads, and continued to hold occasional meetings.

It appears to me then to be demonstrated, that the new Junto was not a mere continuation of the old, incorporated into it by election; and yet the name and the organization were the same. How is this remarkable correspondence to be accounted for? It might at first be supposed that the new Society was one of those Clubs formed by the members of the Junto, as proposed by Franklin, with the same rules, but without being informed of their connection with the parent Club. But to this view there are some objections. One is that the date of this project was 1736, fourteen years before the formation of the new Junto. Another, that the new Clubs had new names given to them, such as the Vine, the Union, the Band; while this bore the old name itself. Another difference is, that it had not, like these branches, a member of the old Club in its number, to connect it with the parent institution.

When the branch Juntos were formed, the original Society was in full vigor and activity, and the new Clubs were established in order to prevent the necessity of increasing the number of the old. But in 1750 circumstances had changed. The members of the old Junto must then have felt that the original objects of the institution had been so far accomplished in regard to themselves, that they need no longer submit to the restraints and labors enjoined by the rules, but might indulge themselves at their meetings in the pleasures of social intercourse. It seems, then, that they did not think it desirable or proper to introduce young members among them, and they ceased to supply any vacancies in their number. But this course would inevitably lead to the destruction of an institution that had proved itself so useful, to which they were warmly attached, and to which it may be supposed they would be ambitious to give perpetuity. How then could they prevent its dissolution? One way was presented to them, and there seems to be but one; and that was to form a Junto of young men, to whom they would give their name and their business, and communicate even their secret laws and regulations; thus making over to them, as it were, all but the privilege of being recognized as the remnant of the ancient Junto, and of meeting as such, from time to time, for social enjoyment. Such a Club was actually formed. It had in it the sons of Franklin and Syng. When, after some time, it was in danger of declining, we know from a letter of Charles Thomson, already referred to, that Franklin urged "the members to exert themselves for its revival," and that in consequence "new members were elected, and the meetings became more regular," so that the new Junto was approved and cherished by the founder of the old.

The identity of name, of number, of day of meeting, of rules of proceeding, of qualifications on initiation, cannot have been the result of chance, and as the old Junto was a secret Society, they can have been received only by direct communication from it. For all these reasons, there can, I think, be no doubt that the Junto of 1750 was formed with the knowledge and approbation of the original Club, and probably at its suggestion. But if this be the case, we can hardly suppose that the name *Junto* would have been given to it, unmodified even by the epithets *new* or *junior,* unless with the intention of making it the legitimate successor of the old company on its retirement. This view of the matter is strengthened by more than one analogous case which has occurred in Philadelphia, where the old members of long-established fire companies have given their name and their apparatus to young men able and willing to undertake the active duties, while the remnant of the ancient company, under the same name, has continued to meet as a social club.

We shall see, as we proceed in our narrative, that this Junto was one of the great branches from which our present Society is formed, and it is gratifying therefore to be able to establish, through it, our descent from the memorable Club of 1727, which contained so many distinguished men, originated so many important measures, and will ever hold a cherished place in the annals of this community. Such a connection was certainly claimed by the earliest members of this Society. Thus Dr. William Smith, in a eulogium pronounced before it on the decease of its illustrious first president, Franklin, says, in speaking of the original Junto, that "after having existed forty years, and having contributed to the formation of some very great men, besides Franklin himself, this Society became at last the foundation of the American Philosophical Society." The credit here given to the first Junto is, in every view of the question, too exclusive; but the reverend orator would hardly have ventured to make the claim, to any extent, on such an occasion, unless he had himself supposed and believed it to be understood by his audience, that the Junto of 1750 had succeeded to the honors, as well as to the name, of that of 1727.

The minutes of the Junto are, many of them, in such detail as to enable us to be present as it were at their proceedings, and to gratify a curiosity which is the greater, as we know that the sessions of the new Club must have resembled those of the old, in the manner of conducting business, in the kind of questions proposed, in the nature of the discussions, in all—except the presence of Franklin.

By a perusal of these minutes we see that the questions were very various, comprising subjects of natural philosophy, natural history, moral science, history, politics. The discussions in general show great sagacity, and no inconsiderable extent of knowledge.

Among the questions discussed are such as the following:

How may the phenomena of vapors be explained: of hail, of storms, the fall of the barometer before rain?

Why do the tides rise higher in the Bay of Fundy than in the Bay of the Delaware? Why are tides at a distance from the equator higher than those near it? What becomes of the water constantly flowing into the Mediterranean?

Is there an essential difference between light and heat? Between the electric fluid and elementary fire?

Questions of a political character were frequently introduced; such as the following:

What form of government contributes most to the public weal? Which was the first that prevailed among mankind? Can any one suit all mankind?

Is there danger of the depreciation of our present paper currency?

Is it consistent with the prerogatives of the crown, and the security of the people's privileges, that the executive powers of government over any territory should be made hereditary, and transferable in the family of any subject?

Should the truth ever be punished as a libel? Should a grand jury hear evidence on both sides?

But I find that I am presenting to the audience a catalogue of queries, which is likely to run into too great a length; and I must therefore content myself now with saying that questions of equal interest were offered on many other speculative and practical subjects; and that the members followed the example of Franklin, by also discussing projects for the advantage of the city: such as the education of children at the public expense; the establishment of public baths; means of supplying the poor with firewood at a moderate cost,—and the like.

In the records, a summary of the debates is given by the secretary, but it is remarkable that the names of the speakers are never mentioned.

One of the exercises required of the members was what is called in the minutes a *declamation*. It seems to have been an unwelcome task, for fines were frequently imposed for its neglect. Although the term *declamation* was used, it consisted simply of a written paper, read by the author. Thus in February 1760, there is the following minute:

"Charles Thomson read to the Company a declamation, in which, from matters of fact joined with probable conjecture, it was endeavored to account for the *gossamer*, or those filmy threads which are sometimes seen to cover whole fields in a dewy morning, especially in the autumn, and to show that they might be the threads or webs of the flying spider, weighed down by the particles of water they collect, from the condensing of vapors, upon the sun's withdrawing his beams."

On this minute it may be proper to remark, that we must not infer from the use of the term "flying spider," that Mr. Thomson was ignorant of the fact that no spiders have wings; for the same name is still applied to such of these insects as have the faculty of floating through the air attached to fibers of their own spinning.

From the 22d of October 1762, until the 25th of April 1776, an interval of three and a half years, no minutes of the Junto are known to exist; yet it is certain that meetings must have been held, for when the minutes are resumed we find that six new members had been added; and besides, during the two first of these years, Dr. Franklin was in Philadelphia, and it must therefore have been to this time that Charles Thomson refers when he says, in his letter to Franklin of Nov. 1762, "From some conversation I had with you, some few of us exerted ourselves to revive it (the Junto) again, new members were elected, and our meetings became more regular."

During this interval Owen Biddle and Isaac Paschall had been appointed "to revise the laws, and to make, a few alterations in them," and the new code was adopted on the 30th of May 1766. It differs very little, however, from the old. Yet from this time a new spirit seems to have been infused into the body. Their views became gradually enlarged, and their ambition excited. A time had arrived when it was thought that they might aim at higher things than the mere mutual improvement of a limited and secret club; when they might increase their number, open their doors, and extend their influence for the promotion of knowledge over the American colonies.

The first step in this course was to add to their list of members; and this they did without regard to the old limit of twelve. To the engrossed copy of the laws in the minute book thirty names are subscribed. At the close of the year (Dec. 13, 1766), rules were adopted for the admission of nonresidents as corresponding members, so that the bounds of the Society could now be extended to the utmost limits of the Provinces, and even into Europe.

By such a step as this, the association lost its character of a club, and accordingly, at the same meeting, it abandoned the name of Junto, which had been used by itself and the parent institution for nearly forty years, and adopted the more ambitious title of "The American Society for promoting and propagating Useful Knowledge, held at Philadelphia."

The Junto thus became a *Society*, took the broad name of *American*, and announced as its aim the promotion of useful knowledge. To act up to these great pretensions was not an easy task, and immediate success was not to be looked for. We ought not, therefore, to be surprised at finding that for many months little was accomplished by the Junto Society that can be said to correspond with its new claims. This failure was painfully felt by some of the members, and particularly by the most zealous of them all, Charles Thomson. Schemes were proposed, from time to time, for placing the Society really in the station to which it aspired, when, on the 1st of January 1768, Thomson reported, "that he had, as far as he was capable, collected the sense of the Company on the subject, and committed it to writing, with such thoughts as had occurred to himself on reconsidering the matter." This paper he then read under the title, "Proposals for enlarging this Society, in order that it may the better answer the end for which it was instituted, namely, the promoting and propagating of useful knowledge." It presents an extended and interesting view of the numerous subjects of inquiry calling the attention of the cultivators of science in America. Every department of knowledge seems to be touched upon, but a special reference is always had to immediate utility, and agriculture is the favorite topic. The paper concludes in the following words:

"The spirit of inquiry is awake, and nothing seems wanting but a public Society, such as the American Society is now proposed to be, formed on a plan to encourage and direct inquiries and experiments, collect and digest discoveries and inventions made, and unite the labors of many to attain one grand end, namely, the advancement of useful knowledge and improvement of our country. As Philadelphia is the center of the colonies, as her inhabitants are remarkable for encouraging laudable and useful undertakings, why should we hesitate to enlarge the plan of our Society, call to our assistance men of learning and ingenuity from every quarter, and unite in one general noble attempt, not only to promote the interests of our country, but to raise her to some eminence in the rank of polite and learned nations."

These proposals were printed and distributed at the time, and subsequently the substance of them was embodied in the preface to the first volume of our Transactions.

The year 1768 continued to be one of great activity in the American Society. Large additions were made to their list of fellows and correspondents, and among them were Dr. Franklin himself (then in England) and other men of great distinction. The proceedings were no longer those of a debating club, but of a learned society. Papers were read, some of which are introduced into our printed Transactions; machines and inventions were submitted for examination; communications were received from foreign correspondents; premiums were awarded; donations were received, and a cabinet formed.

On the 23d of September a code of laws was adopted, suited to the new position assumed by the Society, which was now called, by a slight alteration of the former name, "The American Society, held at Philadelphia, for promoting Useful Knowledge." On the fourth of November it made its first election of officers, and the list presents to us the following distinguished names:

President, Dr. Benjamin Franklin.
Vice President, Samuel Powel.
Secretaries, Charles Thomson, Thomas Mifflin.
Curators, Dr. John Morgan, Lewis Nicola, Isaac Bartram.
Treasurer, Clement Biddle.

We have now to welcome the return to the fields of science of another of our parent institutions. The American Philosophical Society, after struggling some years without success, had fallen into total inactivity. The failure showed that in 1743, the time was not ripe for founding a scientific institution in the colonies, and that in making the attempt, Franklin was before his age. And where else, indeed, should we expect to find him? He was born to be a leader among his fellow men, and his place was always in the advance. If this enterprise, however, was seen by its founders to have been premature, the noble views that gave it origin were not abandoned by them, and they awaited with confidence the coming of a season when a renewed effort on their part might prove more successful. In a young and thriving community, ages of improvement pass over rapidly, and the period thus looked for was not long delayed. The members remaining in Philadelphia of the old Society, reduced to six in number, thought (to use the expression of one of them) that "they saw their way clear for its revival," and they accordingly took measures for this purpose. At what time they first reassembled, we do not know; but it appears that in Nov. 1767, they elected four new members, and in the following January no less than forty-four, including men of the highest rank and most distinguished talent in the colony. From the 19th of Jan. 1768, minutes of the proceedings were regularly recorded, and are preserved in the archives of our Society. These minutes show that the revived institution began its new career with great advantages. Even before the election of its regular officers, the governor of the province, John Penn, consented to become its patron; the use of the council chamber at the State-house was granted for its meetings; the college rooms and apparatus were also put at its command, "whenever the members should choose to meet there, or have any experiments performed before them." It was considered, in fact, as more immediately connected with the aristocratic or proprietary party, and therefore enjoyed the special countenance and aid of the men in power.

On the 9th of February (1768), the Society elected the following officers:

President, The Hon. James Hamilton.
Vice Presidents, Dr. William Shippen, Dr. Thomas Bond.
Treasurer, Philip Syng.
Secretaries, Rev. Dr. William Smith, Rev. Mr. John Ewing, Dr. Charles Moore.

The first president, James Hamilton, had preceded John Penn as governor of the province, and was a man of wealth and influence; but it does not appear that he ever attended the meetings of the Society after his election.

On the 22d of March 1768, the first scientific communication was made to the Philosophical Society, and it now stands as the first paper in our series of Transactions. It is entitled, "A Description of a New Orrery, planned and now nearly finished by David Rittenhouse, A. M." The instrument described has been an object of wonder for three quarters of a century. It is not a philosophical toy, employed for the purpose of giving a general notion of the movements of the planets round the sun; but a mechanical ephemeris, made to exhibit the relative positions of the heavenly bodies at any past or future epoch of time. To construct it, required a man who should be at the same time a scientific astronomer and a skillful mechanician, and such a man was Rittenhouse. Two of these orreries were actually made; one for the college at Princeton, the other for the college at Philadelphia.

Throughout the year, the meetings of the Society were kept up with great spirit, and many communications were made, which are to be found in the printed Transactions. But by far the most important proceedings of the Society were those that had relation to the transit of Venus over the Sun's disc, that was to occur on the 3d of June 1769, and to which the attention of astronomers throughout the world was anxiously directed. On the observations of this transit depended the more accurate determination of that great astronomical element, the sun's parallax; from which is deduced the distance of the Earth from the Sun, and thence the absolute distances of all the planets, their relative distances being already known by other means. The greater interest was attached to this phenomena as it is of very rare occurrence. Indeed it had been before then but twice observed. The first time was in 1639, when Venus was in its ascending node; a transit which, after an interval of two hundred and thirty-five years, will again be seen in 1874, and is the next to be looked for. The occurrence of this transit was foreseen only by one individual, named Horrox, who lived near Liverpool. All other astronomers were ignorant of its happening, and therefore failed to observe it. Horrox had computed the transit from improved tables of Venus, corrected by his own observations; and yet this is considered one of the least of his astronomical performances, though he died at the early age of twenty-two.

The next transit of Venus occurred in 1761, the planet being then in its descending node; and it was carefully observed in different parts of the world, and important conclusions drawn from it as to the sun's parallax. It was known, however, that its second recurrence, which was to take place in 1769, would be under more favorable circumstances; and at the time of the new organization of the Junto, and of the revival of the Philosophical Society, the attention of learned men throughout the civilized world was anxiously directed to the great approaching astronomical phenomenon. To show the importance attached to it, I need only mention that the first of the celebrated voyages of Captain Cook, of which the story is so familiar to us all from our childhood, was made by order of the British government, for the special object of carrying out astronomers to observe the transit at the island of Otaheite, where it was known that the whole transit, which would occupy about six hours, could be seen; an advantage enjoyed by few other observers.

The attention of the Philosophical Society was first formally called to this subject, by a written application from one of its secretaries, the Rev. John Ewing, afterwards for many years pastor of the First Presbyterian Church in Philadelphia, and provost of the University of Pennsylvania. In his letter he states that, "having gone through the calculation and projection of the next transit of Venus, he found that the beginning, and a great part of it, would be visible at Philadelphia, if the weather should be favourable." He then speaks of the importance of multiplying observations of it in different parts of the world, and concludes by saying, "I would humbly propose to this Society, that effectual provision be made for taking the said observations in this city, which is the more necessary as such another opportunity will not occur for more than a century."

This recommendation was referred to the Committee for Natural Philosophy and Astronomy, "to consider the proposal, and make some estimate of the probable expense of preparing for and making the observations" and at the meeting of June 21st, the Society took active measures, by appointing a committee to make the necessary preparations, and to observe the transit at Norriton; and another committee, to erect an observatory at Philadelphia, and make preparation for ascertaining the latitude, and for observing the transit; and the Society agreed to defray the expenses of the observations in both places.

This was a noble undertaking for a Society just beginning its career; but it was at the same time a very difficult one, and it was soon found that it could not be accomplished without aid. Accordingly, a memorial was presented in September to the Assembly, then in session, praying their assistance. It is pleasing to find that this application was acted upon favorably and promptly; for, at the next meeting, Secretary Bond reported that the Assembly had voted a sum not exceeding one hundred pounds sterling, for purchasing a reflecting telescope, &c. I may here mention that this instrument, made by Nairn, with a Dollond's micrometer, was afterwards procured in London by Dr. Franklin.

We have now traced the progress of both the American and the Philosophical Society, to the close of the year 1768, and we have seen that they were engaged, with great zeal and activity, in the same pursuits, and under nearly the same organization. It was impossible, therefore, that they should not both feel the importance of being united, and accordingly negotiations were early set on foot for this purpose.

To the American Society belongs the honor of making the first overtures. On the 26th of January, the question was discussed, "whether, since the two Societies had the same views, it would not be desirable that they should be united, if, this could be done on an equal footing, and on terms equally honourable to both" and it was "voted unanimously, that such a union would be desirable, and would conduce to the public good, if it could be effected on these terms, but on no other."

After a consultation between some of the leading members of both institutions, this minute was sent, on the 2d of February, with a list of fellows and correspondents, to the Philosophical Society. The course taken by this Society, on the occasion, was a singular one; for they immediately suspended their rule requiring proposals for membership to lie over for one meeting, and introduced all the members of the American Society into their own body by election. This measure was communicated to the members, with an assurance that every thing respecting them had been conducted with the greatest marks of regard, and with the same good disposition, which they had shown, for uniting in the common design for the advancement of useful knowledge; that it had been agreed to postpone the proposal of new members and the election of officers until the following meeting, when they might be present, and give their votes and advice.

This course of proceeding was not relished by the American Society, who did not consider that it brought about a union equally honorable to both parties; and they prepared an answer, in which they state, that although their election into the other Society might be deemed an honor to them as individuals, yet as a Society they could not consider it in that light; and they then go on to assert their claims in such terms as plainly show their feeling of offended dignity. Before this communication was delivered, however, some of the members, apprehending that it might give offence, had a special meeting called, when the answer agreed upon at the preceding meeting was suppressed, and the following brief resolution was substituted for it: "The minute of the American Philosophical Society, of the 2d of February, which declares our election into that Society, being considered, it was unanimously determined that, as it was not on the terms proposed, we are under the necessity of declining the union." This minute was communicated to the Philosophical Society, and no measure with regard to it was taken for some months.

Such a state of disunion, however, between two kindred institutions, interested in the same great objects, and entertaining a sincere regard for each other, was unnatural, and could not be permitted long to continue. The next overtures came, as of right they should, from the Philosophical Society, which, on the 15th of November, appointed their two vice-presidents, two secretaries, and two members, a committee "to concert measures, and prepare the way for a union." This proceeding, being made known to the American Society, was kindly but cautiously received, and a committee of conference was appointed, consisting of the vice-president, two secretaries, two curators, and a member, and instructions were given to them as to the conditions to be required for ensuring perfect equality between the contracting parties. The negotiations were conducted with a degree of diplomatic formality, which shows the importance that was attached to the measure on both sides. The desire of union, however, was sincere, and on the 20th of December, both societies being in session, the terms on which it should take place were mutually agreed upon. They were the following; and it will be perceived how cautiously they are framed with a view to the perfect equality in the claims of the two parties to the treaty.

1st. The united society to bear a name composed of the former two, viz. "The American Philosophical Society held at Philadelphia for promoting Useful Knowledge."

2d. No new members to be proposed until the union takes place.

3d. The first meeting to be held at the College on Monday, the 2d of January 1769.

4th. The officers to be one patron, one president, three vice-presidents, one treasurer, four secretaries, and three curators, to be chosen by ballot at the first meeting, excepting that instead of electing a patron, the governor of the province be requested to act as patron.

5th. A new set of laws to be formed, taking in whatever may be thought proper out of the former laws of both Societies.

6th. Each Society to pay off its debts before the joint meeting, and the new treasurer then chosen to receive the balances in the hands of the other two.

7th. The books and collections of the two Societies to be handed over to the united Society.

8th. In the joint publication of transactions, no preference to be given to the papers of either, but that they be arranged and digested according to their subjects and dates.

9th. That there be a new book for the proceedings of the united Society, and that it be opened with a preface or declaration, stating the circumstances of the union, &c.

The ratification of this treaty was the last great act of the two rival Societies, and at the close of the year 1768, a few days afterwards, their existence as separate bodies came to a termination.

The 2d of January 1769, is an epoch in the history of our institution. On the evening of that day, the united Society (our present Society) held its first meeting, and its first election. From the very equal distribution of all the subordinate offices among the members of the two parent institutions, it would seem that there must have been a previous arrangement of the ticket generally agreed upon; but we have the evidence of the late venerable Bishop White, that, as to the presidency, there was an active contest. The candidates were, Dr. Franklin, president of the American Society, and identified with the popular party in politics, and ex-governor Hamilton, president of the Philosophical Society, and a leader of the proprietary party. At the time of the union the total number of members was 251, of whom about 124 resided in the city and county.

Of this number no less than 89 voted, showing how great an interest was felt in the election. The result was a happy and a proper one. The philosopher triumphed, and Franklin was chosen the first president of a society, in which he already possessed the higher title of "the Founder." The vice-presidents, Dr. Thomas Cadwalader, Dr. Thomas Bond, and Joseph Galloway, Esq., were appointed to wait on the governor, John Penn, and request him to be patron of the Society; and they reported at the following meeting, that he had declined the office. Bishop White has said, that the language which the governor used on the occasion was, "No, gentlemen, I cannot be the patron of a Society whose first president is the greatest enemy of my family." Two years afterwards, his successor, Governor Richard Penn, showed a better feeling. When asked to accept the place of patron, he consented in the most courteous terms, his answer concluding in the following words: "I beg leave to assure you, that I shall not consider the patronage of the Philosophical Society, begun and flourishing in this province, as the least honourable appendage to my present appointment."

The first business of the new Society was one of legislation; and at an early meeting, it adopted a code of laws, which is substantially the same as that by which it still continues to be governed.

But the great scientific enterprise commenced by the Philosophical Society, of taking effectual measures for observing the expected transit of Venus, was now to be resumed by the united body, and measures for this purpose were promptly taken. The means of the Society not being sufficient, aid was solicited from the Assembly, and it was liberally granted, by voting money toward erecting observatories, and giving liberty to place one of them in the State-house square. Suitable temporary observatories were constructed accordingly; one in Philadelphia, the other at the residence of Mr. Rittenhouse, in Norriton township, Montgomery County, about twenty miles northwest of Philadelphia. Measures were also taken for making observations at Cape Henlopen on the Delaware Bay, where a building was found that could be used for the purpose.

Committees were appointed by the Society to conduct the observations at the three different stations. Good instruments were provided. One telescope, with a Dollond micrometer, was procured at London, by Dr. Franklin, with the money voted by the Assembly;

another, of the same character, was sent by Thomas Penn, one of the proprietaries, with a request that after it had been used for the transit, it should be given to the College; where it now is. Other instruments were supplied in sufficient number and of good quality. Careful observations were made at each station, for determining the essential elements of latitude, longitude, and time.

The great day so long predicted, so anxiously expected, so carefully, expensively, and laboriously prepared for, was at hand. An envious cloud might disappoint all hopes, and render all the preparations vain. At the observatories in the north of Europe this actually occurred. How was it in Pennsylvania? The Rev. Provost Smith, who was one of the observers at Norriton, says: "The prospect before us was very discouraging. The first of the month, and several preceding days, had been overcast with clouds and frequent heavy rains. But on the 2d the weather cleared up; and on the 3d, the day of the transit, there was such a state of serenity, splendour of sunshine, and purity of atmosphere, that not the least appearance of cloud was to be seen." He adds, that the sun was so intensely bright, that the colored glasses sent from England with the reflector could not be used, and a deeply smoked glass had to be substituted.

The long-looked-for moment was now approaching. A concourse of the inhabitants of the county, and many persons from the city, had crowded about the observatory, and there was some apprehension that the perfect silence necessary on the occasion might be broken. But so great, says Dr. Smith, was the interest and anxiety of the company during, the critical period, "that there could not have been a more solemn pause of silence and expectation, if each individual had been waiting for the sentence that was to give him life or death."

If such was the anxiety of the mere lookers-on, what must have been the feeling of the observers themselves, when the predicted moment brought with it the sight so long expected—witnessed but twice before since the creation—the *Venus in sole visa* of Horrox! Dr. Rush, in his eloquent funeral oration on Rittenhouse, pronounced before the Society in 1796, mentions a report that "in the instant of one of the contacts of the planet with the sun," our astronomer was affected by "an emotion of delight so exquisite and powerful, as to induce fainting." I am happy to say that the deliberate and detailed record that Dr. Rittenhouse gives of all the phenomena, including both the contacts, is inconsistent with this improbable story. I would not certainly underrate the intensity of his emotions: But he was performing a duty of which he felt all the importance, and "how could he find leisure to be sick?" It is known that, in urgent circumstances, even the epileptic can postpone his fits; and I cannot believe that Rittenhouse, while engaged in observing the transit of Venus, would permit himself to faint.

The observations, at Philadelphia, at Norriton, and at Cape Henlopen, were all successful, and the account of them and of the results to which they led is given in full detail in the first volume of our Transactions. In no part of the world were they more perfect or more important. Writing to Thomas Penn, Dr. Maskelyne, the astronomer royal, says: "I thank you for the account of the Pennsylvania observations, which seem excellent and complete, and do honour to the gentlemen who made them, and to those who promoted the undertaking."

I have dwelt with the more detail on the account of this great work of the Society, because it was its first, and because I believe that the anticipation of the transit, and the necessity of preparing for it, was a principal cause of the revival of one of the parent institutions, and afterwards of the happy union of the two.

One would suppose that this astronomical enterprise would have been labor and honor enough for the first year of the Society's existence. Yet another work was undertaken by it, scarcely less laborious or less expensive. On the 7th of April 1769, the Committee for American Improvements were instructed to take measures for the purpose of ascertaining "the best place for cutting a canal to join the waters of the Delaware and Chesapeake, with the probable expense that would attend the execution of it." This work was prosecuted with great spirit. An appeal for pecuniary aid was made to the Philadelphia merchants, and they answered it with the liberality for which they have ever been conspicuous. Committees were appointed to conduct the surveys and levels; and on the different routes no less than fifteen members of the Society were thus employed. The results of their labors, illustrated by a map, are given in the first volume of Transactions. I will not attempt an analysis of this paper; but will merely mention, that the upper route, for a barge navigation (afterwards begun under the direction of Mr. Henry Latrobe, and unfortunately abandoned) was preferred by the committee; and that "they declined making an estimate what the cost would be to make a clear passage from river to river, judging it an undertaking beyond the abilities of the country." This last plan, however, has, as you know, been since executed; but it was at an enormous cost, which has proved ruinous to the company that undertook it.

Very soon after the union, the Society appointed a committee to revise the papers on hand, and to prepare a volume of Transactions for the press. This committee reported a list on the 18th of August 1769, and a new committee was appointed to superintend the publication.

We can hardly realize, at the present day, how difficult a task was the printing of this volume. A whole year elapsed before the first part, containing the astronomical papers, was finished; and it was not till the 22d of February following, that the Society had the pleasure of presenting to the Library and to each member of the Assembly, a copy of the entire volume, "as an acknowledgment of the grateful sense which they retained of the public patronage and encouragement which they had received from the Assemblies of Pennsylvania." The address on the occasion concludes with the following paragraph:

"As the various societies which have of late years been instituted in Europe have confessedly contributed much to the more general propagation of knowledge and useful arts, it is hoped that it will give satisfaction to the members of the honourable House, to find that the Province which they represent can boast of the first Society and the first publication of a volume of Transactions for the advancement of useful knowledge on this side of the Atlantic; a volume which is wholly American, in composition, printing, and paper, and which, we flatter ourselves, may not be thought altogether unworthy of the attention of men of letters in the most improved parts of the world."

The ordinary proceedings of a learned society are not generally such as can make any figure in a mere narrative like the present.

It is not by joint undertakings, such as those we have been speaking of, that the cause of science is usually advanced; but by individual efforts and solitary labors, of which the results, after being matured in the closet, are at length communicated to the public, and thus made to form part of the common stock of knowledge. It therefore happened, that although the Society went on with great activity and zeal, and without interruption in its labors for some years, yet as its "promotion of useful knowledge" was, for the most part, effected by individual exertions, there are but few marked places in its course to which I can further direct special attention.

There was indeed a great effort made, on the suggestion of the president, Dr. Franklin, to promote the culture of silk, by encouraging the growth of mulberry trees, and establishing a public filature at Philadelphia. Application was made to the provincial Assembly to give their aid to this scheme, but they adjourned without acting upon the case. After this, a society, for this object, was formed under the auspices of the Philosophical Society, and subsequently the Assembly voted £1000 to aid the undertaking. The business was committed to managers, who set up a filature, where, as they state, "silk was prepared and reeled on public account" and specimens were presented to the Society, "in consideration that the laudable design was first set on foot by them." Our present venerable President showed himself a worthy successor of Franklin in this cause, by the great and expensive efforts which he made to promote it, by spreading information before the people, urging Congress to follow the example of patronage set by a provincial Legislature seventy years before, and at last by establishing a filature at his own cost.

The eminently useful science of agriculture attracted a large share of the early attention of the Society, and no branch of it perhaps more than the cultivation of the vine. The first premium awarded—and it was by the American Society before the union—was for a successful effort in this branch of industry; and there is a very long and laborious paper on the same topic, in the first volume of Transactions, by the Hon. Edward Antill of New Jersey. This essay begins with a eulogy on wine, which would hardly be relished by the temperance advocates of the present day, although it does end by the words: "wise and happy is the man that shuns excess, that prudently avoids turning this cordial into a cup of poison, and moderately enjoys the blessing with a thankful heart." A scheme for establishing a public vineyard was submitted to the Society in January 1743; but they informed the proposer of it that though "they highly approved the encouraging of the culture of vines, yet as lotteries were contrary to the laws of the Province, they could not countenance the undertaking." Thanks to the exertions of a member of our Society now present, lotteries are again unlawful in Pennsylvania, and it is impossible now, as it was seventy years ago, to build schemes of public improvement upon a basis of public demoralization.

In the minutes of August 1773, there is a record which is too curious to be passed over without notice. It is the report of a committee, of which David Rittenhouse was the chairman, on the first steam engine erected in America. It was made by Christopher Collis for the purpose of pumping up water at a distillery. We cannot avoid smiling, in this day of steam power, when we find a committee give a favorable report, and declare the "undertaker worthy of public encouragement" because "they saw the engine perform several strokes" though in consequence of its execution "being attempted at a very low expense, it did not continue its motion long."

How different is the state of the mechanical arts at the present time in Philadelphia; where one of the members of our Society has constructed, for a public institution, an engine which for perfection of form, of workmanship, of finish, and of operation, may challenge the world; where another, the first to make a locomotive engine in America, is now, by these productions of his skill, bearing anxious travelers and rich burdens along nearly every iron road in the country; where another has made, for the new steamships constructed for our protection and defense, gigantic engines, which in ancient times would have been deemed a work to be accomplished only by Vulcan and his Cyclopes; where a fourth has been found boldly and successfully to carry American competition into Europe, so that the traveler in England, in France, in Prussia, in Austria, may find himself drawn upon his road by engines that bear upon them the names of William Norris and of Philadelphia.

From the month of March 1774, there was an interruption of some months in the meetings of the Society; and similar interruptions occurred, from time to time, in several of the following years. No one acquainted with the history of the time need ask the reason.

There are duties sometimes required of us, as men and as citizens, before which the pursuits of science, however useful or attractive, must give place; and at the period of this first suspension, the encroachments of the British government occupied the thoughts and engaged the active exertions of all true patriots, to the exclusion of such pursuits as require leisure and a mind at ease.

On a temporary resumption of the meetings in December, the following remarkable note appears in the minutes, in the handwriting of Dr. Benjamin Rush, one of the secretaries.

"The acts of the British Parliament for shutting up the port of Boston, for altering the charters, and for the more impartial administration of justice in the Province of Massachusetts Bay, together with the bill for establishing popery and arbitrary power in Quebec, having alarmed the whole of the American colonies, the members of the Philosophical Society, partaking with their countrymen in the distress and labors brought upon their country, were obliged to discontinue their meetings for some months, until a mode of opposition to the said acts of parliament was established, which they hope will restore the former harmony, and maintain a perpetual union between Great Britain and the American colonies."

It will be remembered, that in little more than eighteen months after this hope of harmony and perpetual union was expressed, the writer became himself one of the signers of that memorable instrument which declared the eternal separation of these American colonies from the mother country.

Indeed the impossibility of a reconciliation soon became apparent. In April 1775, the first blood was spilt at Lexington; in May, Congress assembled at Philadelphia; and in June was fought the battle of Bunker's Hill. During such times as these, a Philosophical Society, bearing and meriting the name of "American," could not be expected to go forward. There were accordingly very few meetings in 1775; but there were two of these worthy of notice. At the meeting in February, Rittenhouse delivered an oration on astronomy. It is one of a series given successively by Dr. Smith, Dr. Rush, Mr. Rittenhouse, Mr. Matlack, Mr. Owen Biddle, Dr. Bond, and several other distinguished members. It is full of ingenious and original thought, and in some parts even reaches the sublime. Another event of this year was the attendance of the president, Dr. Franklin. He returned to America on the 5th of May and presided at the Society for the first time on the 15th of September following; nor did he again take the chair until after his return from France in 1785.

In 1776, the annual meeting in January for the election of officers was the only one held, and from that time the Society did not again come together, until after an interval of more than three years. How, indeed, could it have been otherwise? During part of this time, Philadelphia was occupied by the enemy and during the whole of it, war was raging over the land.

But at this pause my narrative must cease; for I cannot, in compassion, allow myself to tax your patience so much further, as would be required to give even a sketch of what remains untold. On the 5th of March 1779, the Society reassembled, never again to be dispersed or to be interrupted in its scientific pursuits. Here, then, begins a new era in its history, and what remains must be narrated by a new historian.

Among the prominent subjects that will require the notice of this historian, one of the first will be the charter obtained by the Society in 1780, from the General Assembly. It was granted in a very flattering manner, is of the most liberal character, and gave to the Society, for the first time, a legal position, which it still occupies. Another point will be the generous donation of John Hyacinth de Magellan, for establishing a premium to be awarded to the authors of discoveries and improvements—and the rules adopted with regard to these premiums. Another will be the grant by the Legislature, of a lot in the State-house square, for the erection of a hall, the measures taken for putting up the building, the debts and difficulties brought on by this great undertaking, and the final relief from them by the united and liberal efforts of the members, bringing, as they always will, confidence and assistance from the community around.

The historian will also have to speak of the distinguished men of the Society, such as Rittenhouse, Jefferson, Wistar, Patterson (who succeeded Franklin in the President's chair), the two Hopkinsons, the two Vaughans, and so many others, to whose memory we owe a tribute of respect and affection.

But the most exact and appropriate measure of the task that remains, compared with that which has been so imperfectly performed, is presented by the amount of the Society's scientific labors. Up to the time at which my narrative ceases, that is, when the meetings were interrupted by the war of the revolution, there had been but one volume of Transactions printed. Since the reassembling in 1779, there have been published thirteen quarto volumes of Scientific Transactions; besides several octavo volumes of Transactions of the Historical and Literary Committee, and of the Society's Proceedings, as reported by the Secretaries.

What treasures are laid up in this great store-house of thought and observation! The best eulogy that could be pronounced on the Society on this day of its jubilee, would be an analysis of these products of its scientific labors; but I have neither the time, the ability, nor the courage to undertake it. Let us hope, however, that it will not be neglected; and that this great occasion will be taken by the Society, for requiring from such of its members as are fitted for the task, methodical reports of what it has accomplished in the several departments of science.

What is the present condition of the Society? This question is a most important one, and would require, for its satisfactory answer, far more time than is left to us. I will boldly assert, however, that so far as its scientific pursuits are concerned, the Society was never more active, or more successful, and never had more talent, learning, and zeal engaged in its service. There are always to be found, in every community, individuals who mourn over the degeneracy of the times; and such persons may be surprised that I should dare to compare the present days of the Society with those of Franklin and Rittenhouse. I do so, however, with confidence, and I appeal to our recent publications and our present labors to support my position.

Will any one ask me to bring a parallel for the labors of Franklin in electricity? He puts me to a most severe test. Nothing so brilliant will perhaps ever again occur in this science, as the proof of identity of lightning and electricity, or that beautiful effort of genius, the Franklinian theory. But to those who are acquainted with the immense advances which the science has made since Franklin's time, it will be seen how much more difficult it is now, than it was then, to make any additional steps in it, or to explain and generalize the obscure relations found between electricity and magnetism. Yet in three of our latest volumes, there will be found a series of contributions on these united sciences, filled with the most curious and interesting discoveries, and the most sagacious theories; the results of great and continued personal labor and patient thought; and the investigations are still going forward. I cannot, therefore, acknowledge that any evidence of degeneracy is shown in such labors as these of Professor Joseph Henry.

But I am next asked whether successors are found to Rittenhouse in the science of Astronomy, and my interrogator points to the transit of Venus. I answer that such a celestial phenomenon as this will not come at our command, and that none perhaps of the same importance has been since observed anywhere. But that the science in which our Society first distinguished itself, is cultivated with as much zeal and success now, as it was in the days of Rittenhouse, Smith, and Ewing, is fully evidenced in our recent Transactions, which are filled with interesting papers on the subject. But instead of referring you to these, I will call your attention to an astronomical phenomenon fresh in your recollections—the remarkable comet which has just disappeared from the heavens. How very different is the character of the observations and calculations made on this comet, and on those reported in our early volumes. They are carried to a degree of accuracy, not attainable with the instruments and methods of former times. Our fellow members, Messrs. Walker and Kendall, who made the observations at the High School Observatory, have been indefatigable in their calculations to determine the orbit of this comet, and have been obliged to repeat their work, and even change their methods, in consequence of the strange peculiarities which the orbit presents; and the amount of labor which has thus been performed, for the satisfaction of scientific curiosity, is hardly credible. I will add, that the calculations of this orbit were also made by our fellow members, Alexander, of Princeton; Anderson, of New York; Peirce, of Harvard University; Loomis, of Ohio, and several others. I cannot then acknowledge any degeneracy among us in the science of astronomy.

I shall conclude this defense of times present, by one more example of scientific enterprise. It is known that the magnetic needle, whether moving in a horizontal or vertical plane, is subject to frequent irregular changes in its position, indicating corresponding changes in the magnetism of the Earth. Similar changes are also indicated by suitable instruments in the intensity of the Earth's magnetic power. Observations made by Arago and Kupffer, in 1818, at distant stations, showed that these small variations were not confined within moderate limits, but took place nearly at the same time, at places very distant from each other. This proved that their cause was more widely extended than had been supposed, and excited the greatest curiosity and interest as to its nature and real limits. To determine these points, it was necessary that observations should be made, at the same time, in different places, spread over as large an extent of the Earth's surface as possible. The first arrangement for this purpose was organized by Humboldt in 1826, at eleven different stations, chiefly in the Russian empire. Gauss, having greatly improved the instruments, induced the philosophers of Germany, in 1825, to form a Magnetic Association, which numbered eighteen observatories under its direction. The simultaneous observations, at places widely separated in the continent of Europe, showed that the movements of the needle were affected by causes, not limited to a narrow locality, but as extensive as the chain of observatories itself. In 1837, the British Association, aided by the Royal Society, and patronized by the government, succeeded in greatly extending the plan and means of observing; and finally a system was organized for making simultaneous observations, at as many stations as possible, in the four quarters of the globe. This great enterprise is known by the name of the *Magnetic Crusade.*

Of the observatories, there are three in the United States; one in Philadelphia, one at Cambridge (Mass.,) and one at Washington. The first of these is attached to the Girard College, and went into operation in June 1839; but the observations were made under the direction of one of our most active officers, and their expense was borne by subscriptions of the Philosophical Society, its members, and their friends, until December last, when support from this quarter became no longer possible, and the observatory was closed. I am happy, however, to be able to say that it is now again in full operation, under means supplied by the intelligent liberality of the present Secretary of War, upon the recommendation of the Chief of the Corps of Topographical Engineers, our fellow-member, Col. Abert.

Now, I could not have selected a better example of the zeal and industry, with which scientific researches are at present pursued among us, than is presented in this great undertaking. It had continued, before the interruption in December last, three and a half years in operation; and during the whole of that time, day and night, bi-hourly observations were made of all the instruments, including a complete set for meteorology; and in the term-days, as they are called, magnetic observations are made every two minutes. Besides this, on particular occasions, and for particular objects, the number of observations is frequently increased. But, excluding these extraordinary occasions and the term-days, there are four hundred and seventy-two regular observations made every day, of which two hundred and six are of the magnetic, and two hundred and sixty-six of the meteorological instruments. This is an immense amount of labor; yet the superintendence of the whole of it, the arrangement and collation of the reports, and the thousand cares and duties devolving upon the chief of such an observatory, are performed without remuneration, though not without personal expense. I cannot think that all this is consistent with the empty cry of a decline of scientific zeal among us in modern days; nor can I think that any degeneracy is shown from the character of Franklin in the Director of this observatory, his great-grandson Alexander Dallas Bache.

How many other proofs could I not readily present that the true spirit of philosophy is not wanting among our members! It would be an easy and a grateful task to mention names and works; but I know not which I could omit. On the whole, it is impossible for any candid person to look upon the present scientific feeling of our Society, without complacency and approbation.

But is there no dark side to this picture? Have I, in this view of the present condition of the Society, to speak only of successful labors and prosperous enterprises? Would that such were the case!

Would that it were in my power to close this address, without the necessity of announcing that misfortune has found its way into our quiet halls! Unhappily our institution has not escaped its share of the troubles that have affected the whole community. An opportunity presented itself, full of tempting plausibility, by which it was believed that the Society could greatly better its condition, could secure ample space and a convenient arrangement for its overflowing and still rapidly increasing library, better accommodation for its members, and an increase of income; while, at the same time, it might save another institution, in which it had always taken much interest, from the destruction which threatened it. The Society yielded to the temptation, and became the purchaser of the Museum property. The step proved to be a most unfortunate one. A fall in the value of property, which no sagacity could have foreseen, has by an indirect operation, which it would be out of place to explain here, deprived the Society of its purchase, and left it involved in debt. We are now in the midst of the difficulty. How is it to end? What account is our future historian to give of it? It will be such—we dare not doubt that it will be such—as is worthy of the Society, and of the community in which it is placed. He will say that the members disdained to receive the blow in listless and cowardly submission; that they did but rouse themselves under it to still greater activity, like men worthy of their social descent; that they united themselves together, as with one heart, to meet the crisis; that they did not hesitate to submit to personal sacrifices in order to save their institution; that they then gathered their friends around them, appealed to their fellow citizens for aid, and had their call generously answered; that they succeeded, as such a course always must succeed; that the city, the state, the country, would not allow the most ancient and the most active of its learned societies to fall; but that it recovered from its losses, resumed its wonted position, and went on prospering and to prosper.

The Society having determined further to celebrate its hundredth anniversary by a special and public meeting of its members and correspondents; such a meeting was held at the Hall of the

Society, on Friday 26th of May, which was extended, by adjornments, through eight sessions. It was attended by Dr. Chapman, Dr. Patterson, and Dr. Bache, *Vice-Presidents*; Mr. Kane, Prof. Bache, Dr. Dunglison and Mr. Fisher, *Secretaries*; Dr. Hays, Mr. F. Peale, and Mr. Wetherill, *Curators*; Mr. Ord, *Treasurer and Librarian*; Mr. Steen Billé, Chargé d'Affaires of Denmark, Mr. Peter, H. B. M.'s Consul, Dr. Warren and Mr. Borden, of Boston, Gov. Dickerson, Professors Henry, and S. Alexander, of New Jersey, Dr. Ducatel, of Maryland, Mr. Nicollet, of Washington, Lieut. Wilkes, U. S. N., Major Bache, U. S. A., Mr. Baldwin, Mr. Bancker, Dr. Bell, Rev. Dr. Bethune, Col. Biddle, Mr. T. Biddle, Prof. Booth, Mr. Boyé, Mr. Breck, Mr. Campbell, Mr. Carey, Dr. B. H. Coates, Dr. Condie, Mr. Cope, Prof. Cresson, Mr. Dobson, Rev. Dr. Dorr, Mr. Ellett, Dr. Emerson, Mr. Fraley, Prof. Frazer, Mr. H. D. Gilpin, Mr. T. Gilpin, Dr. Goddard, Prof. Griscom, Dr. Hare, Dr. W. Harris, Mr. Heinbel, Dr. Horner, Mr. Justice, Prof. Kendall, Mr. Kuhn, Dr. La Roche, Mr. Lea, Rev. Dr. Ludlow, Mr. Lukens, Dr. M'Euen, Mr. C. M'Euen, Dr. Mease, Dr. Meigs, Mr. Merrick, Dr. Mitchell, Mr. Morris, Dr. Morton, Prof. Nulty, Mr. T. B.. Peale Mr. Rawle, Prof. Reed, Mr. Richards, Mr. Roberts, Prof. H. D. Rogers, Prof. Sanderson, Mr. Saxton, Mr. Seybert, Mr. G. W. Smith, Mr. R. C. Taylor, Mr. Trego, Mr. Tyson, Mr. Vanderkemp, Prof. Vanuxem, Prof. Vethake, Mr. Wagner, Mr. Walker, Mr. T. I. Wharton, and Dr. Wood, of Philadelphia, *members*; by delegates and correspondents from various learned societies, and by numerous strangers and citizens, as visitors.

SPECIAL MEETING.

First Session, 26th May, 1843, 10 *o'clock, A. M.*

Dr. Chapman, Vice-President, in the Chair.

The meeting was opened by the chairman with congratulations to the members on the occasion of their assembling, and expressions of welcome to the strangers who honored it by their presence.

Professor Bache, from the subcommittee, consisting of himself, Dr. Dunglison, and Dr. Ludlow, submitted the program of the meeting, so far as it had been matured, and indicated the manner in which the various communications that had been received would be presented to the notice of the Society.

Letters relating to the celebration were read from—

The Northern Academy of Arts and Sciences, Hanover, N. H.,—the Rhode Island Historical Society,—the Connecticut Historical Society,—the Georgia Historical Society,—the Dean of the Faculty of Jefferson Medical College,—the Dean of the Faculty of the Pennsylvania College of Medicine,—the Hon. John Quincy Adams,—the Hon. Albert Gallatin,—the Hon. Daniel Webster,—the Rev. Dr. Samuel Miller, of Princeton, N. J.,—the Rev. Dr. Samuel F. Jarvis, of New. York,—B. Silliman, Jr., Esq., of New Haven,—A. Litton, Esq, of Nashville, Tennessee,—Samuel H. Smith, Esq., of Washington, D. C.,—Professors D. H. Mahan, Hitchcock, Bartlett, and Baily, of the United States Military Academy, West Point,—Professors Peirce and Lovering, of Harvard University,—Chancellor Frelinghuysen, of the University of New York,—Professors W. B. Rogers, R. E. Rogers, and Courtenay, of the University of Virginia,—

President Wayland, of Brown University,—John Pickering, Esq. of Boston,—Horatio Hale, Esq., Boston,—W. J. Andrews, Esq., Secretary of the Boston Athenmum,—Professor Olmstead, of Yale College,—Dr. T. Romeyn Beck, Albany, N. Y.,—Hon. James Madison Porter, Secretary of War,—Professor Loomis, of Western Reserve College,—Dr. Charles Caldwell, of the Louisville Medical Institute,—John L. Stephens, Esq., of New York: Dr. William Darlington, of Chester County, Penn.,—Col. J. J. Abert, of the United States Topographical Engineers,—Dr. Mower, United States Army,—Lieut. Gilliss, United States Navy,—Dr. J. F. Ducatel, of the University of Maryland,—Professors Anderson and Renwick, of Columbia College, New York,—Professor Draper, of the University of New York,—W. C. Redfield, Esq., of New York,—Dr. Warren, of Boston,—James Espy, Esq.,—Dr. John Locke, of Cincinnati,—Professor Strong, of Rutgers College,—George Bancroft, Esq., of Boston,—and the Rev. Professor Potter, of Union College.

A letter was received and read from the General Secretaries of the British Association for the Advancement of Science, dated 31st March, 1843, inviting the Society to participate in the next meeting of the Association, to be held at Cork on the 17th of August next.

A note was presented from the Principal of the Pennsylvania Institution for the Instruction of the Blind, inviting the Society and its guests to visit the Institution. The invitation was accepted, and half past four o'clock in the afternoon was fixed upon for the visit.

A letter was read from Professor Rümker, of Hamburg, containing observations on the comet of 1843.

Professor Henry, of Princeton, presented a communication "On Phosphorogenic Emanation," and illustrated, by numerous diagrams, the experiments which he had made on the subject.

It has long been known, that when the diamond is exposed to the direct light of the sun, and then removed to a dark place, it shines with a pale bluish light, which has received the name of phosphorescence. The effect is not peculiar to the diamond, but is common to a long list of substances, among which the sulphuret of lime (Homburgh's phosphorus) is the most prominent. It is also an old fact, mentioned by Canton, that the phosphorescence is excited by exposing the substance to the light of the electrical discharge.

About three years ago, M. Becquerel, of the French Institute, repeated the experiment of Canton, and discovered the remarkable fact that the phosphorescence is excited in a very feeble degree, or not at all, when a plate of glass or mica is interposed between the spark and the sulphuret of lime, although the effect is not apparently diminished when a plate of rock crystal or one of sulphate of lime is similarly interposed. Or, in other words, he found that substances equally transparent do not equally well transmit the exciting cause of the phosphorescence. Hence the old explanation of the glowing of the diamond, namely, that it is owing to the light which has been absorbed and is again given off in the dark, could no longer be admitted; and Becquerel inferred from his experiments, that the exciting cause of the phosphorescence was due to an impression made on the lime by a radiation from the electrical spark, differing essentially from light, and to which he gave the name of the phosphorogenic emanation.

Biot afterwards made a series of experiments on the "permeability" of different substances, in reference to this emanation as it exists in the beams of the sun; and still later, Matteucci, the celebrated Italian experimental philosopher, has investigated and extended the same subject. The younger Becquerel also has published a memoir on the constitution of the solar spectrum, including its phosphorogenic properties. From the notices of the labors of these *savans* on this subject, as they were adverted to by Professor Henry, it appears that all their experiments, with the exception of those before mentioned as made by M. Becquerel, were confined to the solar irradiation, and consequently they do not lessen the importance of a careful examination of the properties of the same emanation, as derived from a different source, and having a different intensity.

The investigations detailed in Professor H.'s communication relate almost exclusively to the emanation as derived from the electrical spark. The apparatus employed in the experiments was a Leyden jar, of the capacity of about half a gallon; and this was charged each time, so as to give a spark between the rounded ends of two thick wires of about an inch in length. The sulphuret of lime was exposed to the light of the spark at different distances, in shallow leaden pans. The first experiments relate to an examination of a considerable number of substances, in regard to their permeability by the emanation. The results of these, which were given at the close of the communication, will serve to corroborate the inference of M. Becquerel, that the exciting cause of the luminous appearance of the lime is not identical with light.

The next experiments are in reference to the propagation of this emanation. Two slits, of about the one-twelfth of an inch wide and an inch long, were made in two screens of sheet brass, and these slits were placed in the same plane with the path of the spark. After the discharge, the sulphuret of lime under the opening was observed to be marked with a narrow line of light well defined at its edges, and shaded off at its ends into a penumbra; the appearance being precisely in accordance with the laws of a radiation in straight lines from a narrow line of emanation.

Experiments were next made to determine whether the radiation of the emanation takes place with the same intensity from every point of the length of the spark, or whether it is confined to the two extremities, or the poles of the discharging wires. For this purpose, the slits were turned at right angles to their former position, so that the emanation could only reach the lime from a single point of the spark. The experiments with this arrangement showed that the radiation is from each point of the line of the spark, but that it is much more intense from the two extremities. This curious result was verified by another arrangement, which allowed the impressions from different points of the spark to be at once compared with each other. Three slits were cut in a thick plate of mica, and this was placed immediately above the lime, so that one of the slits was directly under the end of each wire, and the other midway between the other two. When the discharge was passed over the plate, the lime under the middle slit exhibited a feeble phosphorescence for two or three seconds, and then became dark, while that under the slits at the end of the spark continued to glow for more than a minute. This effect did not appear to be due to the diffusion of the spark at the middle of its course, since the discharge was from a Leyden jar, and the spark, as is usual, in this case appeared as a single line of light, of the same intensity and width throughout its whole length.

The phosphorescence was excited at a much greater distance than was at first thought possible. In a perfectly dark room, the light was observed, for a few moments, when the pan containing the lime was removed to the distance of ten feet from the point of discharge. The intensity of the light, and the time of continuance, however, diminish very rapidly with an increase of distance.

To determine whether the emanation obeys the laws of the reflection of light, a piece of common looking-glass was so arranged with the path of the spark, the slits in the screens, and the pan of lime, that the angle of reflection could be compared with the angle of incidence: but with this arrangement no impression on the lime could be obtained; the want of permeability in the glass apparently preventing any reflection from the silvered side of the mirror. A plate of polished black glass was next used, so as to get the reflection from the anterior surface: the result, however, was of the same negative character as before. It would therefore appear, that glass neither reflects nor transmits the phosphorogenic emanation, except in a very small degree. When a metallic mirror was employed, a well-defined line of light was impressed on the lime from the reflected emanation, and from the position of this it was found that the two angles were equal.

The refraction and dispersion of the emanation were readily obtained, by employing for the purpose a prism of rock salt, instead of one of glass. The dispersion was shown by the conversion of the narrow line of light, by means of the prism, into a broad band.

The next question was in reference to the polarization of the phosphorogenic emanation; and in obtaining a satisfactory answer to this, several difficulties were encountered. Attempts were first made to polarize the beam by passing it through tourmaline: but it was found that this substance is less permeable to the emanation than even glass or mica. Nichols's polarizing prisms were next employed, but no impression could be made on the lime through two of them; and since the emanation is not reflected by glass, and the polarization from polished metal is very feeble, these substances could not be employed in the process. At length an indirect method was adopted, which gave positive results. This was founded on an experiment of Melloni, in his interesting researches on radiant heat. A pile of exceedingly thin plates of mica, prepared according to the method of Professor Forbes of Edinburgh, was placed between the spark and the pan containing the lime, with its plane at right angles to the line joining the middle of the two. In this position of the pile, no impression was made on the lime by the electrical discharge; but when the plane of the pile was inclined to the line just mentioned, so as to form with it the polarizing angle, a luminous spot was excited.

By this change in the position of the pile, the thickness of the path to be traversed by the phosphorogenic beam was considerably lengthened; and yet the permeability was much increased. This remarkable result could only be the effect of the successive polarization of the several parts of the beam, as they passed the several films of mica, and were thus prepared for a more ready transmission by the succeeding films.

After the emanation was found to be polarizable, it was important to determine whether the intensity of the action on the lime would be different, in case the beam were transmitted through crystals in different directions, in reference to their optical axis; but no difference could be observed, when the beam was passed through crystals of carbonate of lime and quartz, parallel and perpendicular to the axis.

From the foregoing results it is evident, that the exciting cause of the phosphorescence of the sulphuret of lime, is an emanation possessing the mechanical properties of light, and yet so different in other respects as to prove the want of identity. That the same emanation also differs from heat is manifest from the fact, that the lime becomes as luminous under a plate of alum as under a plate of rock salt, although these substances are almost entirely different in their property of transmitting heat.

Some experiments were also made to compare the phosphorogenic emanation with the chemical radiation. For this purpose, a sensitive Daguerreotype plate, and a pan of sulphuret of lime, were exposed together to the light of the sky for five seconds. The plate by this exposure was marked with a photographic impression, but little or no effect was produced on the lime. Another sensitive plate and the same pan of lime were similarly exposed to the light of an electrical discharge: the lime was now observed to glow, whereas no impression was produced on the plate. When, however, the plate was exposed very near to a succession of sparks, continued for ten minutes, with a plate of mica interposed, an impression was made.

The sulphuret of lime was also exposed, for several minutes, to the direct light of the full moon, without any phosphorescent effect. A sensitive plate, similarly exposed, according to the statement of Dr. Draper, receives a photographic impression. These experiments, although not sufficiently extensive, appear to indicate that the phosphorogenic emanation is distinct from the chemical, and that it exists in a much greater quantity in the electrical spark, than either the luminous or the chemical emanation.

Professor H. remarked, that in considering these emanations as distinct, he had reference only to the classification of the phenomena; for if they be viewed in accordance with the undulatory hypothesis, they may all be considered as the results of waves, differing in length and amplitude, and possibly also slightly differing in the direction of vibration.

The phosphorescence of the lime may also be excited by exposure to the light of a burning coal; and in this case the emanation is also screened by a plate of mica. It was also found that the magneto-electrical spark from a surface of mercury, excites the luminous condition of the sulphuret; and it has long been known that heat, applied to the bottom of the vessel containing the article, produces the same effect.

To determine whether the phosphorescence could be excited by electro-dynamic induction, a quantity of the sulphuret was placed between two plates of quartz, and a covered copper wire was wound around the whole, so that the lime occupied the axis of a spiral. But when a discharge of electricity was passed through the wire, the lime gave no indications of phosphorescence: the same negative result was also obtained, when the sparks were passed through the bottom of the leaden pan.

It has been supposed that the phosphorescence of the lime is due to the disturbance of the electricity of the mass of the substance, and the continuance of the light to the subsequently slow restoration of the equilibrium. The result, however, of the following experiment would seem to be at variance with this explanation. The lime was thrown into a tumbler of water and sunk to the bottom; but in this situation, when the spark was passed over the surface of the liquid, it became as luminous, and the effect appeared to remain as long, as when the exposure took place in the air.

Professor H. observed that some of the experiments described by him can be repeated with common chalk, although it is not as sensitive as the sulphuret of lime. Some pieces of it, however, become luminous at a considerable distance, and it is not improbable that the chalk cliffs of England are sometimes rendered phosphorescent by flashes of lightning during a thunderstorm.

But the substance that gives the most brilliant light, although the light does not continue so long and is not as easily excited as that from the lime, is the sulphate of potassa. When exposed to the discharge of a jar highly charged at the distance of a few inches below the spark, it glows for a few seconds with a beautiful azure light: and as this salt is not readily acted on by liquids, it was used to determine the permeability of different substances, by placing a crystal of the salt in the liquid to be tested.

It has long been known that the sulphate of potassa often emits flashes of light during the progress of its crystallization; and it is probable that other substances, which are known to emit light under the same circumstances, may also be rendered phosphorescent at a distance by the electrical emanation.

The following is a list of substances that have been examined by Professor Henry, with reference to their permeability by the phosphorogenic emanation.

Transparent Solids	*Transparent Solids,*
Permeable,	Imperfectly Permeable,
Ice	Tourmaline
Sulphate of lime	Mica
Quartz	Flint glass
Sulphate of baryta	Crown glass
Sulphate of potassa	Saltpetre,
Sulphate of soda	Tartaric acid
Borax	Hyposulphate of soda
Citric acid	Copal
Rochelle salt	Camphor
Common salt	
Alum	

Horn (pellucid)
Wax (do.)

Transparent Liquids	*Transparent Liquids*
Permeable,	Imperfectly Permeable,
Water	Muriatic acid
Solution of alum	Sulphuric acid
Solution of ammonia	Nitric acid
Sulphate of magnesia	Phosphoric acid
Nitrate of ammonia	Sulphate of zinc
and all and all weak	Sulphate of lead,
solutions	
	Acetate of zinc
	Arsenous acid
	Ammonia
	Spirits of turpentine
	Alcohol
	Ether
	Oil of aniseed
	Acetate of lead

Dr. Hays made an oral communication "On the Family Proboscidea, their general character and relations, their mode of dentition, and geological distribution," illustrating his subject by numerous specimens, models and drawings.

This family, which embraces four recognized genera—Elephas, Mastodon, Tetracaulodon and Dinotherium—is, he said, a particularly interesting one, in several points of view. It comprises the largest of all the known terrestrial mammalia: the species were numerous: they inhabited the Earth, over a large geographical range, at a former period, in immense numbers: and at the present time the whole race, with the exception of two species only, belonging to the same genus, is wholly extinct. Finally, we are enabled in this family to trace by remarkably close links the connection between the ancient and existing world.

Dr. H., after noticing the general characters of the elephant and mastodon, gave a brief sketch of the progress of discovery of the latter.

Single bones of this animal were occasionally found from an early period after the colonization of this country; but it was not until 1801 that any thing like an entire skeleton was procured. For this valuable contribution to science we are indebted to the zeal and indefatigable exertions of our fellow member, the late C. W. Peale, by whom two nearly complete skeletons were exhumed, near Newburgh, New York. A few years subsequently, one of our former presidents, Mr. Jefferson, engaged General William Clark, so honorably known by his journey to the Pacific Ocean across the Rocky Mountains, to explore Bigbone Lick, for the purpose of collecting animal remains, and furnished the pecuniary means for the undertaking. A very valuable collection of bones was the result of this expedition. From these Mr. Jefferson requested Dr. Wistar to select for our Society a specimen of every thing new and interesting, and to send the duplicates to the French Institute. By some mistake, several of the bones intended for our Society were sent to France; however, the greater number reached us and are now in our cabinet. The specimens sent to Paris, with the drawings and description by Mr. Rembrandt Peale of the skeletons discovered by his father, furnished the principal data by which Cuvier was enabled to develop the history of this extinct animal.

The only very important part of its skeleton not discovered was the cranium. Mr. Peale completed his skeletons by modeling this part after that of the elephant, the nearest allied animal. But in 1838, an entire head was found in Ohio. This specimen, which was figured and described in the eighth volume of our Transactions, was exhibited by Dr. Hays to the meeting.

In 1830 a most valuable contribution to our knowledge of this group of animals was made by our fellow member, the late Dr. J. D. Godman, who communicated to the Society a description of a new genus, which was published in the third volume of our Transactions. The principal characteristic of this genus was the possession of tusks in the lower jaw, whence it was named *Tetracaulodon.* Doubts were expressed at the time respecting this genus; but the discussion that took place on the subject, Dr. H. stated, was too recent to require that he should recall the particulars. But the investigation to which it gave rise was productive of one interesting result that should be mentioned; namely, the discovery of the number of molar teeth possessed by the mastodon, which was supposed by Cuvier to be only twelve, or at most sixteen, but which was proved in a paper communicated to our Society in 1831, and published in the fourth volume of our Transactions, to be twenty-four.

Attention has lately been again directed to these animals, by a large collection of their bones, made by Mr. Koch in Missouri, which was exhibited a short time since in this city, and is now in London. These fossils seem to have deeply interested the English naturalists; three elaborate memoirs in regard to them having been read to the Geological Society. Prof. Grant and Mr. Naysmith, the authors of the two most recent of these papers, fully recognize the genus Tetracaulodon.

Dr. H. next alluded to the general characters and habits of the Dinotherium, the largest of the known terrestrial mammalia. A head of this animal lately discovered is more than three feet long, and as much in breadth. A model of this head was shown. The animal is calculated to have been eighteen feet long.

Some remarks were offered respecting the animal described by Prof. Kaup, under the name of *Mastodon longirostris*. This Dr. H. considered as the connecting link between the Tetracaulodons found in this country and the Dinotherium, and he suggested that it may prove to be a distinct genus. Attention was called to diagrams of the lower jaws of this group of animals: their relations and differences were pointed out, and some observations were made on the natural quinary group, which they formed.

Dr. H. invited attention to the elongation of the lower jaw in the mastodon, as compared with the elephant,—its being armed with a short tusk projecting in a line with the base, in the Tetracaulodon—the elongation of the chin and its downward curvature, with the increased length and curvature of the tusk, in the *M. longirostris*—and finally, the still greater elongation and increased curvature of the chin, and the elongation and increased curvature of the tusk, in the Dinotherium.

The tusks of the upper jaw, large in the elephant, are still larger in the mastodon, whilst in the Dinotherium they do not exist; and

Dr. H. suggested, that when a complete head of the *M. longirostris* is discovered, it will probably be found that the tusks in the upper jaw are of much inferior size to those of the mastodon.

The mode of dentition of this family was next described, and the dental formula of the genera given as follows:

$$\text{Elephant, Inc. } \frac{2}{0}, \text{ can. } \frac{0}{0}, \text{ mol. } \frac{8-8}{8-8}, = 34.$$

$$\text{Mastodon, Inc. } \frac{2}{0}, \text{ can. } \frac{0}{0}, \text{ mol. } \frac{6-6}{6-6}, = 26.$$

$$\text{Tetracaulodon, Inc. } \frac{2}{2}, \text{ can. } \frac{0}{0}, \text{ mol. } \frac{6-6}{6-6}, = 28.$$

$$\text{Dinotherium, Inc. } \frac{0}{2}, \text{ can. } \frac{0}{0}, \text{ mol. } \frac{5-5}{5-5}, = 22.$$

The geological distribution of this family was next considered.

The Dinotherium and *M. (T.) longirostris* inhabited the Earth at as early a period as the Miocene, and other members of the family successively existed up to the present time. Most of the species of mastodon are found in the older Pliocene formations, while the Mastodon giganteum and the extinct elephants existed during the latest portion of this period. The individual of this species, whose head was exhibited, must have become extinct subsequent to the deposit of the materials upon which grow the present forests of Ohio.

Of still more recent extinction must have been the mammoth (elephant), found in 1803 on the borders of the river Lena, encased in ice, the flesh of which was in such excellent preservation, that not only did the wolves and bears eagerly devour it, but the inhabitants actually cut up the flesh to feed their dogs.

Two species of this family, the Asiatic and African elephant, are at present living inhabitants of the Earth; but as their congeners have become extinct under the slow and gradual influence of causes still in operation, they are doubtless destined to the same fate—to have their bones at some future period inhumed with those of the human race and other existing animals, in formations now in progress.

In support of this position, Dr. H. adduced the fact of several animals having become extinct during the brief time embraced within the historical period. The Dodo, for example, numerous formerly in the Mauritius and Isle of Rodriguez, has for more than a century been entirely extinct. The garnet-winged dove, common as late as Cooke's time at Tahiti, was said, on the authority of Mr. T. R. Peale, to be now also extinct. A very remarkable pigeon, not long since abundant at Upolu, one of the Samoan group, has become so rare, that two years since Mr. T. R. Peale could obtain only two specimens; and a year subsequently, though the greatest exertions were made, not a single one could be procured. The circumstance that had caused the extinction of this bird was related. The Bulimus auris vulpinus must also have become extinct within a comparatively recent period. This shell is found abundantly under the soil at St. Helena, but not yet fossilized. It is never found living.

Human bones have never been found fossilized, the creation of our species being too recent for such an event to be accomplished. In the few instances in which they have been found mingled with those of extinct species, the circumstance can be readily accounted for. Human bodies are, however, occasionally entombed by various occurrences; as by the lava thrown out by volcanic eruptions, which have buried whole cities; by landslips and drifting sands; in the chasms caused by earthquakes; and they are sometimes encased in the calcareous, deposits from water. Examples of all these were adduced. The skeletons of those thus entombed will in the course of time become fossilized. The skeletons in the recent formation at Guadeloupe were noticed, and also the human bones found in Travertin, near Santas in Brazil; and some of these bones, with their calcareous incrustation, cemented to fragments of serpula, ostrea, &c., were exhibited.

Dr. Hays concluded with some remarks on the changes that have taken place in organized bodies, corresponding to the change in the condition of the surface of the Earth—on the proofs they afford not only of an intelligence, adapting mechanism to an end, but of successive manifestations of the same contriving intelligence, adjusting the mechanism to the altered conditions under which it was to exist—and on the reverential and exalted ideas this ought to impress on us respecting the wisdom, power, and goodness of the Creator, by whose fiat all things are called into existence and made to perish, and who alone endures forever, and whose years have no end.

The communication was illustrated by entire heads of the Mastodon giganteum and Elephas Asiaticus, several jaws and casts of different species of Mastodon and Tetracaulodon, with series of teeth, casts, diagrams, &c.

In the absence of the author, Professor Frazer gave a succinct analysis of a written communication "On Analytical Trigonometry," by Professor Strong, of Rutgers' College, New Brunswick, N. J.

In this communication Professor Strong has reduced from simple principles, by the application of ordinary algebra, all the formulæ necessary for the computation of circular arcs and their coordinates.

Starting from the simple relations of similar right-angled triangles, he first obtains an expression for the tangent of double an arc in terms of the tangent of the arc itself, $\left(\tan. 2x = \dfrac{2 \tan. x}{1 - \tan.^2 x}.\right)$

Then by changing the form of this equation, taking its hyperbolic logarithm, and developing this by a well-known formula, he shows, that if the tangent of an arc be expressed in terms of the arc, it will be a function of that arc, in which the coefficient is constant, and from the limiting ratio of an arc and its sine that this coefficient is unity. Consequently, the expression becomes,

$$x \text{ (the arc) } = \tan. x - \frac{\tan.^3 x}{3} + \frac{\tan.^5 x}{5} - \&c.$$

Then again, by introducing into his expressions the sine and cosine of the arc (x), and making with the fundamental principle of the hyperbolic logarithm, that, if n be any number, &c., the base of the system of logarithms, then $n = {}^e \log. n$, he obtains, by a series of beautiful and ingenious transformations,

$$\cos. x = 1 - \frac{x^2}{2} + \frac{x^4}{2.3.4.} - \&c., \text{ and } \sin. x = x - \frac{x^3}{2.3} + \frac{x^5}{2.3.4.5.} - \&c.,$$

which enable us always to calculate the values of sine x and cos. x for any finite value of x.

Again, from the same principle, by adopting a different method of analytical reasoning, he obtains the still more familiar formulæ for the sines and cosines of a compound arc, in terms of the sines and cosines of its parts, which are fundamental equations of analytical trigonometry.

Cos. $(x \pm y)$ = cos. x cos. y .$\pm$ sin. x sin. y.

Sin. $(x \pm y)$ = sin. x cos. y $\pm$ cos. x sin. y.

From these he deduces the correlative formula to the tangent of a compound arc, tan. $(x \pm y) = \dfrac{\tan. \; x \pm \tan. \; y}{1 \pm \tan. \; x \tan. \; y}$.

From this expression he proceeds to obtain the value of the circumference of a circle, in terms of the diameter, proving the correctness of his analysis by obtaining the familiar ratio 3.141592 +.

Also, by substituting nx for x, and assuming y = cos. x + sin. x $\sqrt{-1}$, he obtains values of y, by which we are enabled with facility to divide the circumference of a circle into any number of equal parts, and thus to inscribe any regular figure within it.

Professor Bache presented a written communication, by Professor Elias Loomis, of Western Reserve College, Hudson, Ohio, "On two Storms which occurred in February, 1842."

In this paper, Professor Loomis makes an analysis of the phenomena attending two storms, which occurred, respectively, about the fourth and sixteenth of February 1842, and draws his theoretical conclusions in the way pursued by him in a paper on the storm of December 20th, 1836: which was published in the Transactions.

The materials for his inductions, in relation to these first storms, were as follows:—sixty-eight registers containing barometric observations at points between longitude 91° 25' and 52° 38' W, and latitude 47° 34' and 25° 16' N.; fifty-nine registers of the thermometer and weather from military posts; and twenty-two registers of a similar sort from private observers, besides others of the weather merely. To these numerous observations, Prof. Loomis expects yet to be able to add others.

The principal phenomena of the storms are represented on charts, so as to show distinctly the limits of each. Thus, the regions of the United States, where at an assumed epoch the sky was clear or clouded, and where rain or snow was falling, are indicated by different colors. The direction and force of the wind are shown by arrows, and lines of equal pressure and of equal temperature of the air are drawn. There are five charts representing the principal epochs of the storm of Feb. 16th, and seven showing those of Feb. 4th. By the aid of these, Prof. Loomis makes a review of the phases of the storms, weaving into the narrative additional observations of the fall of rain and snow, and the direction and height of the clouds.

He next proceeds to investigate the proximate causes of the phenomena under several distinct heads.   I. The oscillation of the thermometer.   Both these storms, as well as that of Dec. 1836, were accompanied by a considerable rise of the thermometer. During the storm of December and that of February 16th, the thermometer rose 20° above its mean height for the epochs, and in that of February 4th, 30°. Prof. Loomis, after examining the different causes that may have produced this rise, fixes upon the transfer of air from a lower to a higher latitude as the chief; and having already shown that there was such a transfer, brings facts to prove that there is sufficient information to explain the observed results.

The fall of the thermometer, which succeeded the storm, is in like manner traced to the transfer of air, from a higher to a lower latitude.

II. Causes of the rain.   The cause attributed by Prof. Loomis, in the storm of Feb. 16th, is the configuration of the surface of the United States between the Atlantic coast and the mountains, by which moist air from the ocean was raised to a higher elevation, thereby cooling it, and by which it was carried to mingle with a colder upper current from the west or south-west. This rise, however, may in his view be produced by various circumstances.

III. The motion of the wind.   Professor Loomis gives his views of the cause of the winds observed, as follows: "That an easterly wind should spring up on the morning of February 15th, in the region of Ohio, was the necessary result of the greater weight and density of the air to the eastward. But a westerly wind at the same time prevailed a little beyond the Mississippi river. These two winds were partially opposed, and from their opposition, the air between them was elevated somewhat above the surface of the earth. Being cooled by diminished pressure, its vapor is condensed; a portion of it falls as snow, and the remainder forms clouds, which expand and cover the surrounding country. The condensation of the first vapor develops heat, which diminishes the specific gravity of the surrounding air, thereby causing a more decided tendency toward the storm, which increases the precipitation and the development of heat, so that the storm increases in violence as it continues. In the region of greatest condensation, the rise of the temperature was probably greater than at the surface of the earth. As the result of this rarefaction, the air swells up above its usual height, and flows off in every direction, carrying with it the cloud already formed, and causing the barometer to fall steadily as the storm continues to rage."

Professor Loomis investigates the amount of effect attributable to this cause, and makes the application to this particular storm— shows that the violence of the storm must increase—that in meeting the general current from the westward, prevailing in the higher regions of the atmosphere, the center is carried eastward and that there must be a general motion of the lower or surface wind inward, with a tendency to circulate "against the sun"; all these deductions agreeing with observations. He next examines the questions—what caused the first formation of cloud, and why should a storm, when once organized, cease; concluding, in regard to the first point, that each storm begets its successor, the clouds observed on the 15th of February resulting from the atmospheric disturbance by a preceding storm; and in regard to the second, finds the necessary check in the influence of the cold northwest wind, which, in the storm of February 16th, was observed to flow in more rapidly than the center of the storm advanced eastward.

A similar investigation is made of the circumstances of the storm of February 4th, which was one of several centers, and was rendered remarkable by the occurrence of a violent tornado in the northeastern part of Ohio, In this tornado, an inward motion of the air was observed with a circulation against the sun, and is described by Prof. Loomis in the 43d vol. of Silliman's Journal.

IV. Oscillation of the barometer. Professor Loomis argues, that though local changes in the density of the air are the chief causes of the fluctuation of the barometer, nevertheless these oscillations are propagated according to the law of waves, and are felt much beyond the limits of the original disturbing cause: thus the barometric depression occurring with the storm of February 4th, extended considerably south of the region of rain, or even of cloud.

Professor Loomis next proceeds to certain generalizations, treated under the heads of direction of the wind and progress of the storm. Considering the most simple cases (normal cases) of centripetal and rotary storms, or in which the wind flows inwards in all directions, or gyrates about a center, Professor Loomis concludes that neither of the storms now described belongs to either class; that indeed it is doubtful if the motion of the wind over a large portion of the Earth's surface ever conforms strictly to either supposition, but that it frequently partakes of both the alleged motions. Several cases of storms are referred to in support of this position. Defining the center of a storm to be the point where the greatest barometric depression is found, Professor Loomis traces both the storms under discussion, in a general northward and eastwardly direction, but varying both in rate and direction within limits which he assigns.

The conclusions to which Professor Loomis arrives are as follows:

"The following, then, is my view of the origin of such storms as I have been investigating. This generalization will probably include the greater part of winter storms, but will require some modifications when applied to summer showers. Imagine a time perfectly clear, when the wind is from the west, with the barometer and thermometer at their mean height. This may be regarded as the normal state of the atmosphere, and the whole body of air, from its upper limit to the surface of the earth, is moving on harmoniously in one direction. How is rain produced in such an atmosphere? The first requisite seems to be a change of direction of the lower stratum of air. This appears, in winter, to be frequently the effect of a preceding storm. The prevalent westerly current, being temporarily checked in its progress by a violent storm, soon acquires force sufficient to break down all opposition. It supplants the rarefied air of the storm, and not only restores the barometer to its mean height, but the momentum of the excited mass carries it considerably above the mean. This excess of pressure causes a reverse current a little to the westward of a violent storm; and hence we sometimes have a long series of violent storms succeeding each other at nearly equal intervals; and hence, also, a violent storm, succeeded by an unusually high barometer, affords ground for expecting a second storm within one or two days. But this explanation will not apply to all cases; for then, if the barometer should ever settle down to its mean height all over the globe, we never could have another storm. The case here supposed is not likely ever to happen; but even if it should, we cannot admit the consequence attributed to it. Admit such a case to occur, and the sun's heat would be competent to generate a new storm.

"Different portions of the earth's surface absorb the sun's rays in unequal degrees, and afford unequal quantities of moisture for evaporation. The result is, we find bodies of air in close proximity of unequal density, arising from unequal temperature or humidity. Either case would be sufficient to cause a deflection of the lower stratum of air from its normal direction. Suppose then we have the mass of the atmosphere pursuing its wonted course from west to east, while a stratum of a mile or so in height next the earth's surface blows in some different direction; if this direction be from south to north, then the current must be cooled in its progress by change of latitude. This effect may be aided by the inequalities of the earth's surface, and by friction upon the upper stratum of colder air. At the surface of the earth, when the temperature is probably five or ten degrees above the dew point, no remarkable effect may follow. But at a certain elevation, the air is always saturated with vapor. A very slight reduction of temperature causes cloud, and its density and extent will be proportioned to the energy of the causes in operation. If the wind should blow from the north it might happen that no cloud would be formed; but if the direction should be easterly, being partly opposed to the normal current, some portion of this mass would almost necessarily be elevated from the earth's surface, and being cooled, its vapor would be condensed. The first stage of this process then is an abnormal current at the earth's surface, the second is the production of cloud. At this stage, the sky is covered with a veil which checks radiation; the thermometer rises above the mean from this cause, and also from the heat liberated in the condensation. This only adds to the energy of the first abnormal current. More cloud is thus formed, and presently the particles of water having acquired sufficient size, fall rapidly to the earth. The wind being southerly, the thermometer rises. A portion of the atmosphere being thus unusually heated, and loaded with vapor, while the upper limit of the atmosphere remains nearly invariable, the barometer necessarily falls. Thus these causes might continue to operate a long time, acquiring energy by their own action. A limit, however, is soon attained. The rarefaction thus produced creates a tendency in the surrounding colder and heavier air to rush in and occupy its place. Moreover, if the wind be at all easterly, as is usually the case, it partially obstructs the progress of the normal current. This temporary retardation but gives it accumulated energy, and it is soon reinstated with unwonted violence. When the rarefaction is considerable, this rush of air upon the last half of a storm is not generally in the precise direction of the upper current, but more northerly, this air being the denser, and our southerly wind is supplanted by a violent northwester. We have thus a great rarefaction and elevation of temperature under a south or southeast wind with rain, extending over a large territory. This may be called the third phase of the storm, although it differs from the second only in intensity. There is now a general rush of heavier air to fill this void. This rush is chiefly from the north; but an independent cause, that which imparts direction to the upper current, would give us a west wind. Under these two forces the resulting current is chiefly northwest, but everywhere upon the borders the tendency will be inward. The air thus flowing inward towards a central area, forces upwards the warmer air which rises in the middle, and being cooled by elevation, discharges a greater quantity of rain. The currents moving centrally from every point of the compass interfere with each other and pursue their routes spirally inward. We have thus a species of rotation, which in the center of the storm may have a destructive violence, as at Mayfield, February 4, 1842. This is the fourth phase of the storm, and is the case of a violent storm fully organized. This west or northwest wind carries the storm off from a fixed locality, and the storm is transferred necessarily to points further and further east. But this action cannot continue indefinitely. There is a cause in operation that will soon terminate its violence. This westerly wind travels more rapidly than the easterly. The rarefaction at the center of the storm is a cause that acts equally upon both winds. But the one is opposed to the upper current, and the other nearly coincides with it. Hence the one is accelerated and the other retarded. The result is that at successive points farther and farther east, the same storm, after the northwest wind has begun to blow with great violence, has a less duration, the thermometer rises to a less height, the barometer has a smaller oscillation; and thus at a point far eastward, the oscillation becomes nearly extinct; and the only peculiarity observed in the wind is a stronger westerly current succeeding a calm. This is the fifth and final phase of the storm."

Prof. Loomis adds in conclusion:
"It appears to me that if the course of investigation with respect to the two storms of February 1842 were systematically pursued, we should soon have some settled principles in meteorology. If we could be furnished with two meteorological charts of the United States daily for one year, charts showing the state of the barometer, thermometer, wind, sky, &c., for every part of the country, it would settle forever the laws of storm. No false theory could stand against such an array of testimony. Such a set of maps would be worth more than all that has been hitherto done in meteorology. Moreover, the subject would be well nigh exhausted. But one year's observation would be needed. The storms of one year are probably but a repetition of those of the preceding. Instead then of the Guerrilla warfare that has been maintained for centuries with indifferent success, although at the expense of great self-devotion on the part of individual chiefs, is it not time to embark in a general meteorological crusade? A well-arranged system of observations spread over the country, would accomplish more in one year, than observations at a few insulated posts, however accurate and complete, continued to the end of time. The United States are favourably situated for such an enterprise. Observations spread over a smaller territory would be inadequate, as they would not show the extent of any large storm. If we take a survey of the entire globe, we shall search in vain for more than one equal area which could be occupied by the same number of trusty observers. In Europe there is opportunity for a like organization, but with this encumbrance, that it must needs embrace several nations of different languages and governments. The United States then afford decidedly the most hopeful

field for such an enterprise. Shall we hesitate to embark in it? Or shall we grope timidly along as in former days? There are but few questions of science that can be presented in this country to the same advantage as in, Europe. Here is one where the advantage is in our favor. Would it not be wise to devote our main strength to the reduction of this fortress? We need observers spread over the entire country at distances from each other not more than fifty miles. This would require five or six hundred observers for the United States. About half this number of observations is now registered in one shape or another, and this number by suitable efforts might probably be doubled. Supervision is needed to introduce uniformity throughout, and to render some of the registers more complete. Is not such an enterprise worthy of the American Philosophical Society? The general government has for more than twenty years done something, and has lately manifested a disposition to do more for this object. If private zeal could be more generally enlisted, the war might soon be ended, and men would cease to ridicule the idea of our being able to predict an approaching storm."

This communication was elucidated by numerous charts, projected on a large scale, from drawings prepared by the author.

The reading of it having been completed, Dr. Hare made some observations on the several theories of storms that have been presented by meteorologists. He adverted to Mr. Espy's theory, which he considered inadequate to explain the phenomena of the tornado, referring them himself to the action of electricity.

SPECIAL MEETING.

Second Session, 26th May, half past seven o'clock, P. M.

DR. CHAPMAN, Vice-President, in the Chair.

Mr. Samuel Breck read a communication, entitled, "A Historical Sketch of Continental Paper Money, Part II."

This paper was a continuation of one read by Mr. Breck before the Society in June 1840, in which he gave the history of the paper currency of the Revolution, and showed its agency in securing the independence of the United States. In the paper now presented, he proceeded to demonstrate that the nonredemption of the continental money operated on the people of that period as nothing more than a moderate tax; that Congress never repudiated it; and that the omission to provide for its regular liquidation was not without striking precedents in the history of European states.

On the 10th of May, 1775, immediately after the battle of Lexington, Congress prepared its first emission of continental Colonial bills, and on the 22d June, as soon as the news of the battle of Bunker Hill reached Philadelphia, two millions of Spanish milled dollars, so called, purporting to lie for the defense of America, were put in circulation, the confederated colonies standing pledged for their redemption.

In November of the same year, three millions came out in bills of various value, as low as one-third, one-half, and two-thirds of a dollar, and from one dollar to eighty. The colonies were called upon to sink, *proportionally*, a sum of three million. In fixing the proportion to redeem that amount, Virginia was rated the highest,

and stood charged with $496,000
Massachusetts came next, at 434,000
Pennsylvania third, at 372,000
Maryland fourth, at... 310,000
And in the fifth class there are four colonies, all rated
     alike; namely, Connecticut, North Carolina, South
     Carolina, and New York! Each of these stands rated
     at 248,000

Mr. Breck adverted incidentally to this scale, as indicating the relative wealth of the colonies at the beginning of the war, and remarked that New York, whose capital was then unoccupied by the enemy, is rated at little more than half of Massachusetts, while Boston was in the possession of the British.

Before the close of 1775, three millions more were issued; but on the emission of a few millions in addition early the next year, difficulties began to arise. The bills were sometimes refused, confidence was weakened, and depreciation followed. Then came from Congress and the committees of safety-threatening resolutions, denouncing the refractory. It was the first serious emergency, and required prompt relief. Patriotic men, who had the means, stept forward to redeem the bills at par: some of them exchanged as much as a thousand pounds in silver for a like sum in paper. But the remedy was only temporary; for in May 1776, five millions were again emitted, and in the autumn five millions more. Although some specie was imported, it could not avail against such profuse issues. Credit, already on the wane, continued to sink. Nor could Congress ever after fully restore it. The states did not respond to their call for aid.

The power of taxing was virtually denied, by its shackled conditions, in the articles of confederation; and paper continuing to depreciate, an attempt was made, in imitation of the mother country, to raise a revenue by the establishment of a national lottery. The first trial was a failure; for the scheme, which was to sell tickets, *for specie*, at twenty dollars, and pay the prizes in treasury notes, bearing four percent interest, did not induce many to adventure; so that no other resource was left for the prosecution of the war, but a fresh emission of paper money. The people, however, refused to sell their produce for it at par, and Washington was authorized to seize the supplies for the army wherever he could find them, and imprison those who rejected the bills offered in payment.

Having delegated power to compel the circulation of their bills, by military force, Congress on the 27th December, 1776, sent forth another sum of five millions. This was followed up, in the early part of 1777, by further efforts to support the credit of their bills; for which purpose a declaration was published, asserting that they ought to be paid *in full*, fulminating anew against those who impaired their credit, by raising, as the resolution says, the nominal value of gold and silver; and calling upon the states to punish, by forfeitures and penalties, all those who refused to sell their lands, houses, and goods, for continental paper money at specie value.

The whole amount of paper money issued during the war was about four hundred millions of dollars; but the collections made by the continental government in various ways cancelled from time to time the one-half: so that the maximum of circulation at no one period exceeded two hundred millions. Nor did it reach that sum, until its depreciation had compelled Congress to take it in, and pay it out, at the rate of forty paper dollars for one in hard money.

It kept nearly at par for the first year, during which period only nine millions were issued; an amount about equal to the specie then held in all the colonies. And when used in that moderate way, it passed with very little depreciation; but soon after, when the emission increased rapidly, it fell proportionately in value, going on from year to year in its downward course, until Congress, as we, have seen above, fixed the scale by law at forty for one. But million following million in quick succession lessened its exchangeable rate from day to day to the agio of 500, and then 1000 for one; when it ceased to circulate.

Congress had exchanged some of the notes at forty for one, by giving the holder loan office certificates at par, and offered to redeem the whole in the same way, at 1000 for one, when they had sunk to that price. But those very loan office, and other certificates of debt, bore in market no higher price than two shillings and sixpence on the pound, or eight dollars for one; so that very few availed themselves of that offer.

Public securities of similar character, but bearing various names, such as loan office certificates, depreciation certificates, final settlements, &c., were given also to the public creditors who had demands for moneys lent, supplies furnished, services rendered, &c., and these together constituted the Congressional debt at the end of the war. They consisted of obligations or bonds, bearing interest at six per cent, and were entirely distinct in character and tenor from the money bills, which bore no interest, and were used altogether as currency. The value of those certificates in market, as I have already said, was not more than seven or eight for one, until the adoption of the present Constitution in 1789, when they were funded, and rose to par.

In the Journal of Congress of the 29th April, 1783, an estimate is given of the whole revolutionary debt; except the paper money; and it stands thus:

1. Foreign debt to France and Holland $ 7,885,085
2. Domestic debt, in various certificates, as above   34,115,290

                                     $42,000,375

The foreign, bearing interest at four and five
    per cent, and amounting to $ 369,038 06
The domestic at six per cent, and amounting to 2,046,917 04

                                     $2,415,956 10

When the Constitution, by which we are now governed, went into operation, Alexander Hamilton, the first Secretary of the Treasury, added to the domestic debt the claims held by several states against the national exchequer, to the amount of $21,500,000, and then funded the whole; putting a part on interest at six per cent immediately; postponing a part without interest for ten years, then to bear six per cent; and the remainder bearing immediate interest at three per cent.

The arrears of six years' interest were added; which, with some other unsettled claims, made the whole debt amount to ninety-four millions.

Accustomed as we are, at the present day, to the large expenditure of the federal government, we may well be astonished at the economy of the first year of Washington's administration, when the civil list was estimated at 208,000 dollars, and the war department at only 137,000, including even the Indian supplies.

The establishment of a revenue to pay the interest on the debt in 1789, was equivalent to an increase of capital, by bringing that debt to par, of nearly one hundred millions, the greater part of which was held by our own citizens. This was the cause of immediate prosperity, and of the brilliant career that continued for many years after. Every dollar of this aggregate debt was honorably paid.

Mr. B. adverted to the difficulties by which the statesmen of the Revolution found themselves embarrassed on all questions of revenue and finance. They had the best disposition, he said, to pay that currency, and they professed to have the ability so to do. They felt themselves bound in honor to discharge, at their full value, bills emitted by themselves, and bearing on their face a solemn engagement to redeem them in Spanish dollars, or the value thereof, in gold or silver. To do this, however, required a season of tranquility; but the country was invaded by land and by water; it required power to levy taxes, and this was denied them; it required the industry of peaceful times to enable the people to contribute; but the war, in constant activity, baffled every attempt at regular employment. Congress had not even a choice of evils; they had no alternative. One source of revenue only was, at their command, and that was the emission of bills of credit. The very necessity of the case forded them to misuse and abuse it; for even in its depreciated condition, paper money offered facilities so attractive, that the great men at the head of affairs, always intending to redeem it, were glad to find the people willing to receive, at the current exchange, that which could be so easily and liberally supplied.

"Who," said a member, during a debate upon this subject, "will consent to load his constituents with taxes, when we can send to our printer and get a wagon load of money, and pay for the whole with a quire of paper?" And with wagon loads, thus cheaply obtained, they carried on the campaigns of the two years, 1778 and 1779, keeping an army of thirty or forty thousand men in the field; issuing paper, to the amount of sixty-three millions for the former year, and seventy-two millions for the latter; and thus, with an active printing press, and a few commissioners, hired by the day or by the job to sign the bills, ways and means were found to defray almost the whole expense of the civil list, the army and navy, and contingencies. There was, indeed, a little hard money passing through the treasury. The exact sums received in both those years, having been officially reported to Congress, stand recorded on their journals. If it were not attested in this authentic shape, it would be difficult to believe it. The aggregate of gold and silver received into the treasury for the year 1778, was only seventy-eight thousand six hundred and sixty-six dollars; and for the year 1779, the sum of seventy-three thousand dollars: so that the whole machinery of government was carried on, for two entire years, as far as concerned the agency of specie, with one hundred and fifty-one thousand six hundred and sixty-six dollars. So small an expenditure in metallic currency, shows the powerful agency of paper in the belligerent operations, at that critical period; performing as it did, in spite of counterfeits and depreciation, the office of hard money.

This handful of solid coin, which, in gold would weigh only seven hundred pounds, and might be put into a wheelbarrow, was all that came, as we have seen, into the public chest for two years; and we may not be surprised at government being so chary of it, as to refuse General Washington's demand of a small share, to pay a part of the bounty to enlisted soldiers. In denying him, they declared that the precious metals must be kept for the commissaries of prisoners, to be used where paper would not pass.

Congress was driven to the abuse, which ruined the credit of paper money, by the illiberal terms of the confederation, and the reluctance of some of the states to impose taxes. It was authorized only to recommend—not to legislate—and it failed in almost every appeal for pecuniary aid. It was even denied by the single veto of Rhode Island the establishment of an impost of only five per cent on imported goods, which after great difficulty and delay had been ratified by all the other states. Unanimity being a constitutional requirement; that measure, so obviously necessary, so moderate in its amount, so gentle and equal in its operation, was defeated by the negative of the smallest state in the confederation. Nor could the entreaty of Congress, contained in a long argumentative report address to Rhode Island, and drawn up by Alexander Hamilton, James Madison, and Thomas Fitzsimmons, cause that state to retract.

But the last day of the paper currency's usefulness was close at hand. It fell to 1000 for 1 in the early part of 1781, and of course went wholly out of circulation at that period. Two hundred millions of dollars of continental paper lost all their value, and were laid aside. The people who bore the brunt of an eight years' war, and victoriously established independence, sustained without a murmur the whole loss, and voluntarily reduced to utter nothingness the greatest item in the cost of the revolution; and thus waved all claim upon posterity for its payment.

This was, undoubtedly, a severe tax; yet, when examined with care, it will be found less heavy than it seems at first sight. Let us take the largest sum by which the people could ever have been affected—say three hundred millions at twenty for one, which is only half the rate fixed by Congress. Twenty for one on three hundred millions will give fifteen millions of sound money. These fifteen millions having been used as currency for six years, give an annual average of two millions and a half. That sum, among a population of three millions, would not be a poll tax of one dollar; or if the three millions of inhabitants be divided into families of six persons each, making five hundred thousand families, the annual loss per family would be only five dollars! In all probability, the real loss was less to many than this proportion; because the bills passed, with great activity from hand to hand, to their last days, even when five hundred for one; never remaining locked up, nor long withdrawn from circulation. They were divided too into small sums, from one dollar to eighty, and always convertible at the current exchange into every kind of real and personal property; and in their hourly rapid passage left with each temporary possessor the trifling loss only of their daily depreciation.

Mr. Breck reviewed the unavailing efforts, which had from time to time been made by Congress for sustaining the credit of the continental issues, the appeals to the state legislatures and to the people, the attempted levy of a monthly tax, and the perilous resort, in the last emergency, to the acts of limitation of prices, the tender laws, and denunciation. He referred particularly to the legislation, by which Pennsylvania endeavored, in December 1780, to support the credit of its "Island" paper money, and the penalties it denounced against all who should refuse to accept it at par, extending even to confiscation and imprisonment. He spoke too of the "Funding Act," as it was termed, which was passed by that state in March 1785, which denied to assignees the right of receiving interest on a certain class of its public debts; and he remarked with severity on the proposition, which was urged soon after by some political writers, to pay off these debts at the depreciated rates which this very legislation had contributed to induce. He commented at large on the injustice and demoralizing character of all these expedients, and traced their consequences, by numerous references to published history and cotemporary letters, in wild speculation, broad spreading poverty, the destruction of social confidence, and the depravation of morals.

Mr. Breck then alluded to the shutting up of the exchequer for two years by Charles the Second, and the consequent suspension of proceedings for the collection of debts, as given by Burnet; to the refusal of the Spanish crown to recognize the debt of forty-five millions contracted by Philip the Fifth; to the deferred payment of the bills of the Marquis of Vaudreuil on the French treasury in 1782, when, to protect the merchants who had negotiated them from damages, the king retained the bills, and forbade his notaries making any protest. These and other illustrations of permanent as well as occasional delinquency, on the part of foreign governments, were presented by Mr. B. only as showing that the fiscal irregularities of our Revolutionary treasury were not without precedent abroad.

The communication proceeded to sketch the history of the immense speculations in stocks and public securities, which followed the adoption of the funding system and the institution of the first Bank of the United States; and the still more enormous landed transactions of that day, in the monopoly by individuals of millions of acres. It concluded by a comparison of the personal and domestic habits of our people in the later years of the last century, their religious observances, opinions, and feeling, the temper of their political discussions, the impartiality of their tone toward foreign nations, and the diffusive nationality of their patriotism, with those which belong to the Americans of more recent times: and recogniz-ing the severity and truth of the imputations to which the extravagance of public and private speculation has subjected the present generation for a season, Mr. B. denied that an impartial review of the whole history of our country would give reason to apprehend the prospective degeneracy of American character.

Mr. Thomas Biddle, adverting to a part of Mr. Breck's paper, remarked that some nine years after the legislation of Pennsylvania which had been mentioned, that state honorably discharged in full the principal and interest of her debt, for which certificates had been issued. It was classed as follows:

1st. New loan, issued in exchange for evidences of debt of the United States (or Congress) to her citizens.
2d. Militia, issued in payment of the services rendered by her citizens during the war.
3d. Depreciation, issued to make good to her officers and soldiers the loss by continental, money in their pay.
4th. Dollar money, issued by the continental Congress, assessed on and guaranteed by her.
5th. All paper money, issued after the continental money from its depreciation had ceased to be current.

Professor Rogers, of the University of Pennsylvania, read a paper by himself and his brother, Prof. W. B. Rogers, of the University of Virginia, on the phenomena of the great earthquakes which occurred during the past winter, one in this country and the other in the West Indies, and on a general theory of earthquake motion, by which they propose to elucidate several points in geological dynamics.

The essential or characteristic phenomena in every earthquake, as distinguished from those which are occasionally concomitant, are a peculiar wave-like motion of the ground, and a rapid tremulous jar. This was originally stated by the Rev. Jno. Michell, of Cambridge, England, whose generalization the authors confirm by the facts they have collected in relation to the late American disturbances. The undulatory movement, which they conceive to be the *cause* of the vibratory one, extends, usually, to a greater distance than this latter from the center of the earthquake. Observations, gathered from various authentic sources, were cited to prove that the rocking motion is of the nature of a true *billowy pulsation*; evidence deemed conclusive on this head being derived from the earthquake of Conception in 1835; from that of Haiti in May 1842; and from that of the Windward Islands in February 1843.

Professor Rogers next illustrated the manner in which these earthquake undulations advance, and showed, by aid of a diagram map of the United States, the lines along which the shock of the 4th of January last was *simultaneous*, throughout an ascertained distance of at least 500 miles in a N. N. E. and S. S. W. direction.

By a comparison of the times when this shock occurred at the various places affected by, it, the *direction, velocity,* and *mode of advance* of this earthquake, were satisfactorily demonstrated. Thus it was shown, that the area agitated at any given instant was *linear*, being directed N. N. E. and S. S. W., and that this line of synchronism moved to the E. S. E. parallel to itself, and with the enormous velocity of at least thirty-two miles per minute. Numerous observers concur to establish the above inference of the direction taken by the earthquake, by stating that the oscillation itself was from the west to the east. Professor Rogers acknowledged his indebtedness to the Secretary of War and other gentlemen, for aid in collecting much valuable information respecting this earthquake, from remote localities.

He next referred to the recent fearful earthquake of the Windward Islands, which seemed to have its center of violence in the region of Guadaloupe and Antigua. A body of details had been collected in regard to the phenomena of this shock, showing the exact time of its occurrence at each locality, from the Coast of Guyana, through the Eastern Antilles, to Bermuda, from which latter point accurate and useful facts had been promptly communicated by Governor Reid. The length of the region shaken, estimated from Demerara to New York, was about 2300 miles, and the greatest width of the belt, from Bermuda to Savannah, about 770 miles. As frequently the case in earthquakes, the disturbed zone was narrowest in the vicinity of the volcanic portion of the disturbed tract. This earthquake seems to have been generated along a nearly north and south line, running through Martinico, Guadaloupe, and Antigua, to the continent of South America and to Bermuda. The nature of the motion was identical with that attributed to earthquakes generally, only differing from the Mississippi shock in the greater intensity of the action. Accompanying the characteristic undulation and jar, occurred numerous parallel fissures in the earth, which repeatedly gaped and closed, while steam, warm water, and hot sulfurous vapors, found their escape. These phenomena are frequent concomitants of violent earthquakes.

This earthquake was nearly *simultaneous* throughout the extended zone, which embraces the Eastern Antilles, Guyana, and Bermuda, though it came later to the coast of the United States by an interval of twenty-nine or thirty minutes. Its velocity of transmission was very nearly twenty-seven miles per minute. Thus simultaneous, or nearly so, along the north and south line passing through the volcanic axis of the Windward Islands, it was also approximately simultaneous, when felt some twenty-nine or thirty minutes later in a N. N. E. and S. S. W. belt, embracing the Atlantic cities of the United States, from Savannah to New York. This may be accounted for, by conceiving the earthquake to have been propagated from the axis in which it originated in a dilating elliptical form, and the synchronal line of the Atlantic cities to have coincided with the northwestern side of the ellipse, opposite the end of the generating axis or fissure.

Passing, next, to the theory of earthquake motion, under which they propose to unite these facts and generalizations, the authors of the communication attribute the wavelike motion of the surface in earthquakes to an *actual pulsation*, or system of waves, in the molten matter beneath the Earth's crust, occasioned by a sudden linear rending and immediate collapsing of the crust from excessive upward tension, with explosive escape of highly elastic vapors. These waves in the internal fluid lava will impart their undulation to the overlying crust, and lead to all the concomitant phenomena of earthquakes. If the oscillation proceed from a very elongated axis of disruption, the periphery of the earthquake will assume the *elliptical* form; but should it originate in a relatively short fissure, or in a mere focal point, like the vent of a volcano, it will be approximately *circular*.

These views of the origin of earthquakes furnish a new argument in support of the doctrine of central heat; since the frequency of earthquakes in almost every district of the globe implies that the internal igneous fluid must be absolutely coextensive with the surface.

As respects the amplitude of the individual undulations, Professor Rogers conceives that it is practicable, in certain instances, to compute it with some approach to accuracy, by more than one method. While the breadth of the crust-waves must vary with the energy of the earthquake, it can be shown to have been, in some of the more violent of these convulsions, enormously great; in the earthquake of Conception probably ten or eleven miles, and in that of Lisbon as much as twenty-five miles.

Professor Rogers concluded by stating, that the lateness of the hour induced him to withhold the sequel of his paper, the design of which was to apply the generalizations which he had presented, to the explanation of the origin of those great flexures of the strata so magnificently displayed in the mountain chain of the United States; and he announced his purpose to resume the subject at some more appropriate season.

SPECIAL MEETING.

Third Session, 27th May, 10 o'clock, A. M.

DR. PATTERSON, Vice-President, in the Chair.

The Secretaries presented a letter addressed to them by Mr. Sears C. Walker, and Professor E. Otis Kendall, of the High School, "On the Great Comet of 1843." This letter is dated High School Observatory, May 27, 1843, and, omitting a few paragraphs, is as follows:

Gentlemen.—We avail ourselves of the centennial meeting of the members of this Society to lay before them, generally, the reasons which induce us to believe that the recent visitor is a comet of short period, only 21 7/8 years, and that it is identical with those of February 1668, and of December 1689. An early suggestion of its identity with that of 1668 was made, we believe, by Prof. Peirce, in a lecture delivered at Boston on the 23d of March last. Shortly before that date, viz. March 20, it appears to have been noticed by Mr. Cooper, of Nice, in a letter to Schumacher, published in the *London Times*. The question of their identity has been discussed by Prof. Schumacher and Mr. Petersen, of Altona. The latter applies Galle's elements to the perihelion passage in 1668, and Prof. Schumacher expresses an opinion in favor of their identity. The subject has been more fully discussed by Mr. Henderson, the Astronomer Royal of Scotland, who, in a letter to Schumacher of April 11th, states that "there appears great probability in favour of the supposition that the late comet, and the one which appeared in 1668, are the same."

Mr. Henderson then gives the elements of the comet of 1668, and a comparison of the ephemeris computed from them, with the places of the nucleus of the comet, as found by Mr. H. on a map in his possession, containing a trace of its path among the stars from March 9th to March 21st, 1668, as seen at Goa. The agreement is quite sufficient to warrant a conclusion of their identity. The first suggestion of the identity of the comets of 1689 and 1843, was made by ourselves in a letter to the editor of the *Philadelphia Gazette,* April 6th, in which, after giving our own elements of this comet, and Pingré's elements of that of 1689, we mentioned "these elements agree quite well with Prof. Peirce's and ours, except the inclination. The observations used by Pingré are pronounced to be good by Olbers, and he expresses confidence in the elements of Pingré. Still the imperfections of instruments and catalogues of stars in 1689, may have caused such imperfections of the observations as to lead Pingré to an inclination of 69°, instead of 39° or 36° as found at present. When we consider that the inclination found by Prof. Peircé and ourselves is derived from an orbital motion of less than 2°, it is manifest that the position of the plane of the orbit, or, in other words, the inclination, must be quite uncertain: The same difficulty must have occurred in 1689, under still more unfavorable circumstances. It is quite likely, therefore, that a modification of the elements of this comet, not greater than those of Halley's comet in its successive periods, would represent the observations used by Pingré, as well as his own elements, or at least within such limits as those to which the errors were liable."

In a communication in the *Inquirer* of the 11th April, we still repeated our suggestion of the sameness of these comets. Finally, in the *Boston Courier* of April 25th, Prof. Peirce published his elements of the comet of 1689, and found an inclination smaller even than that of 1843, with other elements agreeing very well with those of the recent comet. This removed all doubt in our minds of the identity of these comets, and on the arrival of the *London Times* of April 14th, containing Schumacher's opinion confirmatory of Prof. Peirce's of the sameness of the comets of 1668 and 1843, we compared the periods, to see if the comet of 1843 could not be both that of 1668 and 1689; and we found that a period of 21 $^7/_8$ years would answer for all three. We announced this conclusion in a letter dated May 8th, in the *United States Gazette* of May 11th, with an attempt to account for its not being seen except about the eighth period of its revolutions, when it returns to the perihelion at the same season of the year. We also stated that our parabolic elements, which gave an orbit passing through our first and last normal places of March 20th and April 9th, gave the place on the middle date of March 30th too much advanced. We also stated that such was the case of all the good parabolas obtained for its orbit in Europe or America, and mentioned our coincidence in opinion with Encke, that the parabola was not the true orbit, and added, that probably it would be found to be an ellipse of 21 $7/8$ years. We also stated, that an attempt further to correct the parabola for the middle observation, would lead to a paradox such as Encke had encountered in his attempt to complete an orbit, on the presumption that the curve is a parabola. We immediately, with the kind assistance of Mr. John Downes, commenced the computation of an orbit on Gauss's general method, without presuming upon any conic section, but hoping to find an ellipse, and found a double paradox, a comet moving in an hyperbola, and that hyperbola having its perihelion point within the body of the sun. We immediately announced this result in the *United States Gazette* of the 19th April, and invited an expression of opinion from astronomers as to the legitimate interpretation of this result. It was manifest, that if the center of gravity of the comet and tail was moving away in a nonperiodical curve, our favorite opinion of the identity of these three comets, and short period of 21 $7/8$ years, would be untenable. Although we considered the hyperbolic orbit as well as the small perihelion distance to be both paradoxical, we were willing to submit them as genuine deductions from our observations and computations, and leave them to be received as paradoxes, or explained away, as the sequel should show. In so doing, we postponed, for the time, urging our favorite theory of the short period of 21 $7/8$ years. It is true that we had suggested the probable cause of the acceleration of the comet's place for the middle observation, as computed from a parabolic ephemeris, to be owing to the shape of the comet, in the *United States Gazette* of the 6th of April, after pointing out the acceleration of the comet's place for the middle observations, viz.—"The slight difference between the two curves (our parabola and the true path of the cornet) is lost amidst the errors of observation, and the uncertainty whether the central portion or the densest part of the nebulosity corresponds with the actual center of gravity." We were aware that Encke had resorted to this hypothesis to explain the paradox of the acceleration of his comet, previous to his more fortunate suggestion of the resisting medium. In regard to the recent comet, our attention was early called to this source of error by our esteemed correspondent, Mr. E. C. Herrick, of New Haven, who, in a letter addressed to S. C. Walker, on the 29th of March, remarks, "The concentration of light in the nucleus (as seen in the ten feet Clark telescope of five inches aperture) seemed to me, on two occasions, to be considerably nearer the anterior than the posterior part. Once we thought we could detect three dim star-like points, but it was almost impossible to decide with certainty. *Where the tail is so immense, is there not some hazard in assuming the center of the nucleus to be the center of gravity of the whole body?*" We are particular about the dates of these suggestions respecting the center of gravity of the comet and tail, inasmuch as it is found to be a matter of much importance in the sequel. Having fairly, on the 19th and 20th, laid our two paradoxes, viz. the hyperbolic orbit, and the perihelion distance less than the sun's semidiameter, before the public, with some suggestions as to the inferences that would follow from a strict interpretation of this result of calculation and observations, viz. that of the necessity of a rebound, or of the comet's flowing round the sun, we waited for the opinions of our friends, and for further information from the European observatories. We have since received both, and hasten to lay them before you. First, the arrival of the *Caledonia* brought out the announcements from most of the European observatories in Prof. Schumacher's excellent Astronomical Notices of April 22. From these it appears that the comet's nucleus was first seen in Europe, at Nice, on the 14th, and first observed at Rome on the 17th of March. This was five days later than it was observed at several places in the United States, viz. on the 9th and 11th, not to mention Mr. Clark's measures of the distance of the nucleus from the sun on the 28th of February. The latest observation quoted by Schumacher, is that of Encke at Berlin, March 31st. Perhaps it was seen later. We followed it at the High School Observatory till the 10th of April. The conclusion of Encke,

Steinheil, Nicolai, Schumacher, Argelander, and others, that the parabola is not the true conic section for this comet, confirmed the announcement we had made on the 11th April. Encke, who alone of all the astronomers yet heard from, had discussed the question of the particular conic section, had found an hyperbola resembling ours, with the perihelion point just falling outside of the sun. Thus one of our paradoxes, that of the hyperbolic orbit of the observed center of the nebulosity, was confirmed by the only astronomer in Europe, who, as far as heard from, had gone over the same ground with ourselves.

For the other paradox, viz. a perihelion point within the body of the sun, we find the most ample confirmation stated by the European astronomers: This element is thus state by the European astronomers—

Plantamour, Geneva,	0.0045
Arago, Paris,	0.0054
Galle, Berlin,	0.0118
Argelander, Bonn,	0.0072
Nicolai, Manheim,	0.0037
Encke, Berlin,	0.0047 or less.
Do. do.	0.0036
Do. do.	0.0052 hyperbola.
Mean,	0.0057
Do. omitting Galle,	0.0049
Our last result,	0.0041 hyperbola.
Sun's semidiameter	0.0047

Thus it appears that Plantamour, Nicolai, and Encke, on two occasions, had encountered the same paradox as ourselves, viz. that of a perihelion point within the sun. It is also remarkable that none of the orbits, except Encke's hyperbola, suffice to represent the observed path of the center of the nebulosity among the stars.

Hence it appears, from the concurrence of authorities on these subjects, that good observations of the path of the center of the nebulosity, carefully reduced, lead to a hyperbolic orbit, and an approach of centers of the sun and comet as near as their physical qualities will permit.

In this stage of the inquiry, the principal difficulty consists in reconciling these two paradoxes with our favorite opinion of the identity of the three comets of 1668, 1689, and this year, with a short period of 21 $^7/_8$ years. Now it is fortunate, that in the case of our hyperbola the same natural and plausible explanation that does away with the one paradox does away with the other. The true key to the solution of the difficulty is, we are persuaded, the suggestion first made to us by Mr. Herrick, March 28th, and first suggested to the public, by ourselves, in the *United States Gazette* of April 6th, viz. the "uncertainty whether the central or densest portion of the nebulosity corresponds with the actual centre of gravity." We now proceed to state the opinions of our esteemed friends and correspondents on this point. Dr. Anderson, of New York, writes under date of May 19th and 22d, stating, unhesitatingly, that the analogies in favor of the identity of the comets of 1668 and 1689, should lead us to reject the hyperbolic orbit as being unnatural in itself, and wholly irreconcilable with these analogies; and that we should rather regard this hyperbolic orbit, and too close perihelion distance, as the consequence of some error in the data, or in the methods, or in the computations. That there is nothing in the effect of contact of the bodies, or resistance of the comet by the atmosphere of the sun, which could change the character of the conic section from one of a less velocity to one of a greater. From Professor Alexander, of Princeton College, we have received a letter dated May 20th, in which he proposes an explanation of the difficulty, at once simple and natural, and fulfilling all that was required by Dr. Anderson. It is based on the supposed occurrence of the very error against which we were cautioned by Mr. Herrick, March 28th, and which we alluded to in our published letter of April 6th, namely, the error arising from measuring the place of the densest point of the nucleus, instead of the common center of gravity of the nucleus and tail. We give below his letter in full. We have also had placed in our hands, by Professor A. D. Bache, a letter from Professor Bartlett, of West Point, dated May 23d. We give below that part of his letter which treats on this subject, remarking, that we have no doubt that the coincidence in opinions of Mr. Herrick, Professor Alexander, and Professor Bartlett, has taken place without either one having any knowledge that the same idea had occurred to the other two. We would also remark, that the criticism of Professor Bartlett on Arago's parabolic elements and on our own, is just, and confirms our opinion, that no parabolic ephemeris will perfectly represent a series of observations of this comet. We know of only two sets of elements that will give a good ephemeris; the one is Encke's, and the other is ours. Both are hyperbolic and paradoxical. We give them below. The explanation of Professors Alexander and Bartlett, we have no doubt, is the true one. It is plain and natural, and, *a priori*, extremely probable. It will also satisfy the criticism of Professor Anderson, inasmuch as it points out the particular source of the error of the data, which Dr. Anderson supposed must exist somewhere. The explanation is doubly satisfactory for ourselves, since it leaves the way clear for the establishment of the short period, and the identity of the three comets of 1668, 1689, and 1843, and leaves us still a hope of seeing this remarkable visitor in 1865. Moreover, it does away with both paradoxes, and shows, at the same time, that the European astronomers, as well as ourselves, who were led into them, arrived at them in the legitimate and only possible mode of observation and computation.

Professor Bessel, of Königsberg, the greatest living astronomer, and, since Olbers's death, the most experienced and sagacious observer of comets, remarks, in a letter to Prof. Schumacher, dated March 28th,—"This comet seems to have expended the greater part of its nucleus in building up its splendid tail."

We are happy to add the testimony of our friend Mr. Nicollet, in favor of the strength of these analogies, and of the probable return of this comet in 1865, as an inference not to be in the slightest degree shaken by the fact, that a nice discussion of the observations of the apparent center of the nebulosity has led to the two paradoxes already quoted. We hail the favorable opinion of this distinguished traveler, who received the Lalande medal for the discovery and the elements of the comet of the year 1821. We are happy further to add the testimony in favor of the plausibility of the period of $21\,7/8$ years, communicated to us in writing, or verbally, of our valued friends, Alexander, of Princeton; Mitchell, of Nantucket; Gilliss, of Washington; Herrick, of New Haven; Loomis, of Western Reserve; and, nearer home, of Professors Patterson and Bache.

A letter from Prof. S. Alexander to Mr. Walker, dated Princeton, 20th May, 1843, after remarking that the supposition of the comet having actually struck the sun or his envelope, and then *rebounded*, is too violent, to be admitted, except in the absence of all other rational explanation, proceeds to suggest the following as perhaps a plausible solution of the difficulty. "The centre of gravity of the comet of 1843," he says, "was at an unusual distance from that which seemed to be the actual nucleus: this led to an erroneous estimate of the comet's position. As, moreover, the comet, when first observed, was nearly in its perigee, it is altogether possible that the error arising from the cause here suggested, was, at the same time, at its maximum, and that it continually decreased until the comet disappeared. The effect upon the relative position of the apparent and true orbits would consequently be such as is roughly represented in the following diagram—the true or dotted orbit deviating more and more from the apparent, as we retrace it in the direction opposite to the comet's motion, and thus escaping the sun at the perihelion."

Professor W. H. C. Bartlett, of West Point, writes to Professor Bache, under date of 23d May, 1842, as follows:

"The more immediate purpose of this letter is to suggest to yourself and the Society, what has appeared to me a possible explanation of the very great discrepancies between the observations and both the ephemerides computed from M. Arago's and Mr. Walker's elements.

"I suppose that the *apparent* orbit of the comet is different from the *true*, or that the path of the nucleus is not the same as that described by the centre of gravity of the entire mass. To illustrate my meaning, suppose the comet to approach the sun in a parabolic or very elongated elliptical orbit, which will be, by the principles of physical astronomy, the path of the centre of gravity. As the comet approaches perihelion, let it be greatly but gradually elongated in the direction of a line joining the nucleus and the sun, the tail being thrown off in a direction from this latter body; and suppose this to result from the repulsive action of the cometary particles upon each other, in consequence of the heating influence of the sun, in the same manner as the elastic force of vapor is increased by an elevation of temperature. The action being limited to the particles upon each other, the centre of gravity will be undisturbed,

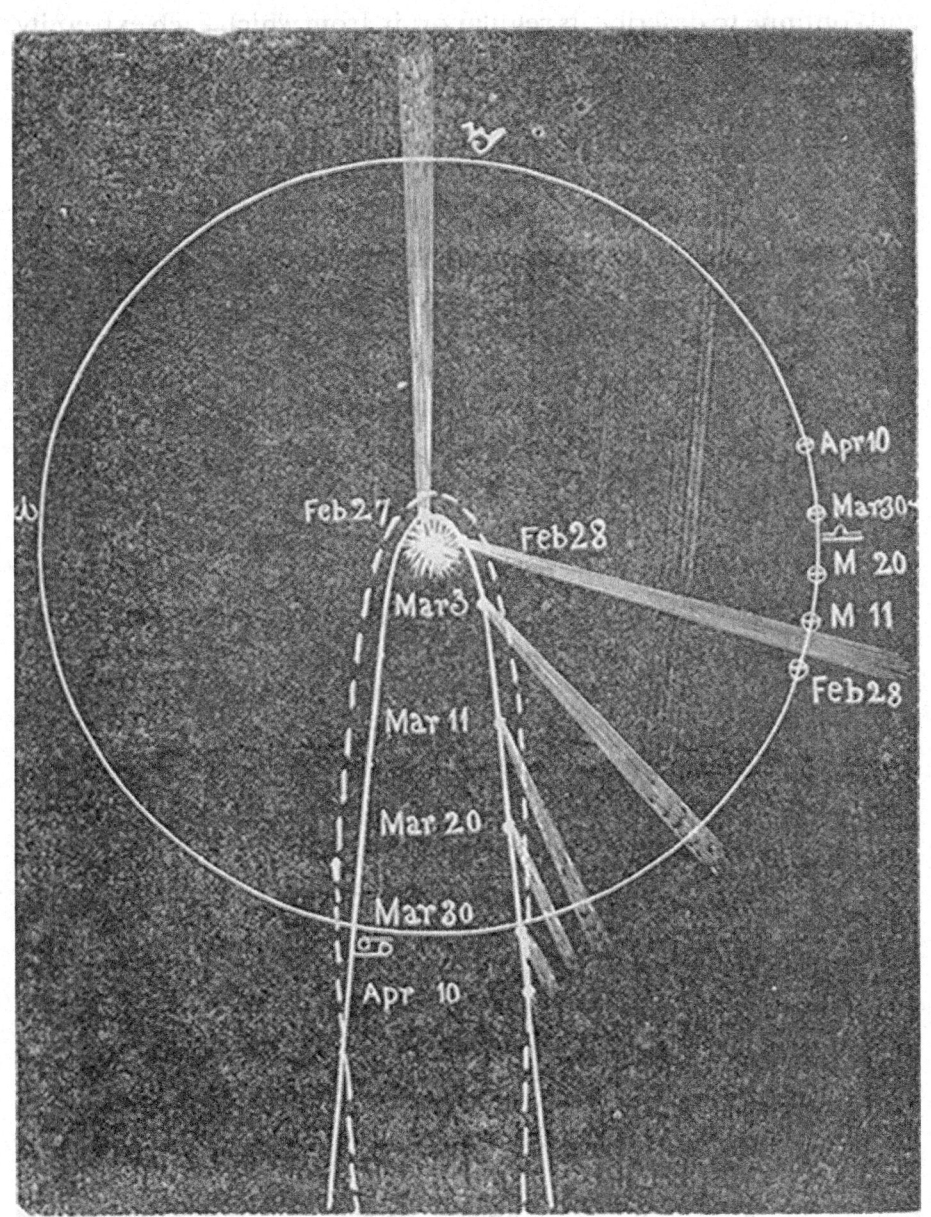

and continue to describe its regular orbit, from which each extremity of the elongation will recede on the line of the radius vector, though in unequal degrees, till it reaches a maximum, resulting from an equilibrium between the elastic force of the cometary medium and the weight of its elementary particles, or the force by which they are drawn toward the center of the mass.

"The expansive action here supposed, would, in the nature of things, be gradual; and hence, before the nucleus, or the *thing observed*, could be totally resolved into a vapor like the tail, and thus disappear, the reverse action would begin, in consequence of the rapid retrocession of the comet from the sun. The disturbed motion of the nucleus being, for a part of the time, from the true orbit, or, that of the centre of gravity, *towards* the sun, the observations, if made at this time, would give a constantly increasing eccentricity, or diminishing perihelion distance; and thus the perihelion itself might be brought apparently within the surface of the sun, while not a particle of the comets matter would touch that body. The observations, if made while the mass of the comet is contracting towards its centre of gravity, would give an increasing perihelion distance till this point is again brought within the nucleus in the depths of space."

We subjoin Encke's hyperbolic elements, and also our own. The latter have been recomputed, after correcting a slight oversight in our calculations, kindly pointed out by Prof. Anderson. Encke's ephemeris agrees closely with his observations. Our hyperbolic elements give an ephemeris corresponding with our normal places within one second of space.

	Encke.	*W., K. and D.*
Perihelion passage,	Feb. 27^d.49778 } m. t. Berlin.	27^d.58939 m. t. Green.
Longitude of perihelion,	279° 2' 29.9 } m.eq. March 0	280° 44' 3.7 m. eq. March 30
Long. asc. node,	4 15 24.9	15 57 3.2
Inclination,	35 12 38.2	34 19 52.0
Eccentricity,	1.00021825	1.00090495
Gaussian angle,*	1°11'49".0	2° 26' 12".1

---

* This angle is the arc whose secant is the same as the eccentricity.

| Perihelion distance, | 0.00521966 | 0.00410367 |
| Daily motion retrograde, | 13″.175559 | 159″.58936 |

It will appear on comparing these elements, that they agree very well, excepting the eccentricity and its secant, the Gaussian angle. This is always the most uncertain element in such investigations. That there can be no error in the process of computation by Mr. Downes and ourselves, is shown by the fact, that the elements reproduce by computation our normal places, after applying the following small corrections, viz:—

March 20.5	R. A.−0″.6	Dec.+0″.7
″ 30.5	R. A−0 .0	Dec.−1 .0
April 9.5	R. A.−0 .6	Dec.+0 .3

These normal places were obtained from a comparison of all our observations with the best ephemeris we could obtain, which was computed from our elements at our request, by Mr. John Downes, the editor of the United States Almanac, and obtained from the average corrections concurring together near the 20th and 30th of March, and 9th of April, for Greenwich mean midnight. These are more correct than the result of any single measure. We give them for the use of astronomers, freed from refraction, parallax, and aberration.

March 20^d.5,	R. A. 46° 4′ 38′.4	Dec. S. 9° 9′ 45″ .5
March 30^d.5,	59 51 1 .2	6 46 32 .5
April 9^d.5,	68 56 41 .6	4 45 35 .7

Let us now consider the period belonging to the mean motion. It is obvious, that if we adopt the explanation of Messrs. Alexander and Bartlett, the mean motion and consequent period of the center of the *nebulosity* observed, and of the real center of *gravity*, must be the same. This is a necessary condition, because they both arrive at the perihelion point at the same instant of time. Now the Earth's sidereal motion in a mean solar day is 3548″.18761. The mean motion of the apparent center of the nebulosity by our elements is 159″.58936. This, if in an ellipse, would give a period for the apparent center of the nebulosity, and consequently for the actual center of gravity, of 22.2339 years.

We now present the argument, *a priori*, derived from analogy, the longitudes being referred to the equinox of the present time.

No.	Comet.	Lon. of Perihel.	Lon. of Node.	Inclination	Perihelion Distance.	Perihelion Passage
1	Comet of 1668	279°.6	359°.8	35°.9	0.0048	Feb. 28.8*d*
2	Comet of 1689	273 .5	346 .5	30 .4	0.0103	Dec. 2.13
3	Comet of 1843	280 .5	348 .5	39 .3	0.0087	Feb. 27.54
4	,,	272 .3	356 .5	36.6	0.0147	27.20
5	,,	262. 7	357 .7	36 .7	0.0541	27.55
6	,,	277 .0	361 .3	35 .7	0.0082	27.45
7	,,	274 .8	359 .1	35 .9	0.0109	27.09
8	,,	274 .5	357 .6	36 .4	0.0113	27.46
9	,,	279 .2	359 .9	36 .0	0.0045	27.46
10	,,	281 .4	365 .9	35 .0	0.0030	27.37
11	,,	275 .5	359 .0	36 .1	0.0104	27.54
12	,,	277 .5	361 .0	35 .7	0.0071	27.47
13	,,	280 .5	364 .6	35 .2	0.0037	27.39
14	,,	278 .8	362 .2	35 .5	0.0054	27.41

No. 1.   By Henderson, Astronomer Royal of Scotland.
2.   By Prof. Peirce, from Pingrés places.
3.   By Prof. Peirce, from his and Mr. Bond's places.
4.   By Messrs. Nooney and Hadley, from Walker and Kendall's early observations.
5.   By Prof. Anderson, from Prof. Bartlett's places.
6.   By Prof. Anderson, from later places of W. and K.
7.   By Prof. Alexander, from his own places.
8.   By Mr. Galle of Berlin,   do.
9.   By Mr. Plantamour of Geneva,   do.
10.   By Prof. Encke,   do.
11.   By Walker, Kendall and Downes, do.
12.   By Argelander,   do.
13.   By Nicolai,   do.
14.   By Laugier and Mauvais,   do.

These arguments seem quite conclusive, and indicate the period of the comet of 21 $7/8$ years, and its identity with some of the many others, quoted by Pingré in his Cometography, as having occurred in the three series of cycles of 175 years (eight revolutions), which precede the respective dates of its recent appearance in 1843.2, its expected appearance in 1864.9 or 1865.0, and in 1886.9. They also completely confirm the observation made by Messrs. Herrick and Bradley, of the eccentricity of the densest portion of the nebulosity in that nebulosity. They confirm the remark we published in the United States Gazette of April 6th. They confirm the coincident opinions of Professors Alexander and Bartlett. They explain away the seeming paradoxes of the hyperbolic motion of the apparent center of the nebulosity, and of the tendency of this fictitious curve to a perihelion point within the sun's surface, while the true ellipse of 218 years' period has a perihelion distance greater than the sun's radius, leaving the comet free to depart and (as we hope) to return about the 1st of January, 1865, to be seen under more favorable circumstances than at this visit.

We conclude by expressing our great satisfaction at the explanation of Profs. Alexander and Bartlett, which, with the computations of the new orbit, by Henderson, for the comet of 1668, and by Prof. Peirce for 1689, have removed the only known obstacle to the admission of the period of $21\,^7/_8$ years, and the elliptic orbit suggested by ourselves on the 8th inst.; accordingly we offer it to the members of the Society, on this their centennial celebration as the probable period of this remarkable comet.

If we admit this hypothesis, and suppose that the perihelion distance was possible, that is, for instance, greater than 0.0047, then we shall find the elliptic elements of the comet's orbit the same as the hyperbolic, omitting the Gaussian angle, and making the eccentricity greater than 0.9994.

The actual elliptic elements may be found on this hypothesis, by assuming the above value of 0.9994 for the elliptic eccentricity, and then giving to the difference between the elliptic and hyperbolic radii vectors the form of a constant quantity, multiplied by the reciprocal of the square of the elliptic radius vector. This constant should then be determined from the series of observations by the method of least squares. The elliptic elements should of course be used in computing perturbations.

The following letter from the same gentlemen, which was communicated to the Society at a subsequent meeting, forms an appropriate, supplement to the foregoing.

High School Observatory, Philada. June 16, 1843.

To the Secretaries of the American Philosophical Society:

GENTLEMEN,—Since writing the letter which was read at the centennial meeting of the Society, we have compared our normal places of the comet on the 20th and 30th of March with the European observations. We have not been able to find any later than the 31st of March, and must still rely on our own measures for the comet's place on the 9th of April. In order to test the normal places for March 20th and 30th, we subjoin the differences therefrom of the European observations referred to the date of Greenwich mean midnight, after rejecting two in all, whose discrepancies from the mean result exceeded fifty seconds of space.

Observation compared with		Correction of normal place, $\Delta\alpha$.	Correction of normal place, $\Delta\delta$.	Date of normal place.
Paris,	March 19.5	+ 7".2	+ 2".2	March 20.5
Geneva,	" 19.5	+ 7 .0	+ 2 .1	
Rome,	" 19.5	+31 .0	+ 5 .0	
Rome,	" 20.5	− 7 .3	+35 .1	
Berlin,	" 20.5	+ 5 .4	−23 .9	
Munich,	" 20.5	+ 3 .4	−21 .4	
Manheim,	" 21.5	−13 .9	−26 .1	
Geneva,	" 21.5	− 2 .6	−28 .3	
Bonn,	" 21.5	+ 0 .5	−28 .1	
Berlin,	" 21.5	− 7 .1	−20 .9	
Munich,	" 21.5	−23 .5	+ 0 .5	
Vienna,	" 21.5		+31 .0	
Mean correction,		−00".0	−6".1	
Manheim,	March 29.5	+ 0".9	−13".9	March 30.5
Bonn,	" 29.5	− 3 .3	−16 .1	
Manheim,	" 30.5	+20 .6	+15 .5	
Berlin,	" 30.5	+ 2 .1	−19 .5	
Berlin,	" 31.5	+33 .3	− 0 .8	
Mean correction,		+10".7	− 7".0	

If we allow to the High School observations the same weight as that of one European observatory, then the normal places of the point of observation of the comet's nebulosity will stand thus, being freed from parallax and aberration.

Greenwich mean time.	Normal place, R. A.	Correc. R. A.	Corrected normal place, R.A.	Normal place, Dec.	Correc. Dec.	Corrected normal place, Dec.
Mar. 20ᵈ.5	46° 4′ 38″.4	+0″.0	46° 4′ 38″.4	−9 9′ 45″.5	−5″.4	−9 9′ 50″.9
Mar. 30 .5	59 51 1 .2	+9 .5	59 51 10 .7	−6 36 32 .5	−5 .2	−6 36 37 .7
April 9 .5	68 56 41 .6	+0 .0	68 56 41 .6	−4 45 35 .7	−0 .0	−4 45 35 .7

Then the corrections of the ephemeris, computed from our hyperbolic elements, will be

$$\text{March 20.5} \qquad \Delta\alpha = -0''.6; \qquad \Delta\delta = -4''.5$$
$$\text{March 30.5} \qquad '' \quad +9''.5; \qquad '' \quad -6''.2$$
$$\text{April 9.5} \qquad '' \quad -0''.6; \qquad '' \quad +0''.3$$

These values are so small that a change in the elements of the orbit of the point of the nebulosity observed, which should reduce them to zero, would be too small to indicate any change in the conclusions already drawn by us from our first normal places. Unless, then, further observations shall be obtained from the southern hemisphere previous to the perihelion passage, we see no way of avoiding the conclusion that the point of the nebulosity observed was moving in a hyperbola, with a mean daily motion of about 160″, which in a curve having a periodical character, would give a duration of a revolution of about 22 years, with elements, as far as we know, identical with those of the comets of 1668.2 and 1689.9.

We subjoin from Pingrés Cometography a list of comets that have appeared at dates when this comet, if it be the same as those of 1668.2, and 1689.9, must have been in a situation to be seen from some part of the Earth. It must be recollected that this comet can never have come to its perihelion in the months of November, December, January and February, without being a conspicuous object in the morning or evening twilight, before or after the passage of the perihelion. In all instances it must have been best seen in the southern hemisphere. We have given nearly all the coincidences in dates. Those which have no (*), nor (?) annexed are coincidences in date. Those marked with an (*) have, besides the coincidence in date, some circumstance, whether of physical appearance or apparent path in the heavens, analogous with the comet of 1843. Those marked with a (?) are probably mere coincidences in date without being the same individuals.

Date.	Periods of eight revolutions preceding the recent appearance.	Single revolutions and mean period.
B. C.　432(?)	13 x 175.00	104 x 21.876
A. D.　268(?)	9 x 175.02	72 x 21.878
442(?)	8 x 175.15	64 x 21.894
617	7 x 175.17	56 x 21.896
968	5 x 175.04	40 x 21.880
1143(*)	4 x 175.05	32 x 21.881
1317	3 x 175.40	24 x 21.925
1493	2 x 175.10	16 x 21.888
1668.2(*)	1 x 175.00	8 x 21.875
1843.2		

Date.	Periods of eight revolutions preceding expected return in 1865.04.	Single revolutions and mean period.
B. C.　60(*)	11 x 175.00	88 x 21.875
A. D.　290(*)	9 x 175.00	72 x 21.875
639(?)	7 x 175.14	56 x 21.892
815	6 x 174.99	48 x 21.874
990?	5 x 174.99	40 x 21.874
1165	4 x 175.23	32 x 21.904
1340*	3 x 174.98	24 x 21.873
1516?	2 x 174.47	16 x 21.809
1689.95(*)	1 x 175.00	8 x 21.875
1865.04		

Date.	Periods of eight revolutions, preceding expected return in 1886.91.	Single revolutions and mean period.
B. C. 213	12 x 174.99	96 x 21.874
A. D. 488	8 x 174.85	64 x 21.856
837	6 x 174.97	48 x 21.871
1012(*)	5 x 174.96	40 x 21.870
1362(*)	3 x 174.94	24 x 21.867
1537(*)	2 x 174.91	16 x 21.864
1886.91		

We have stated the opinion of several able astronomers, that the densest portion of the nebulosity of the recent comet, necessarily selected as the proper point for micrometric measures, was eccentric toward the sun from the real head or center of gravity of the comet, tail, and nebulous envelope. In fact the comet never presented any appearance of a distinct kernel or head, but only a vague and ill-defined nebulosity or cloud, gradually condensed toward the center, or, according to Messrs. Herrick and Bradley, toward a point nearer the sun than the center of the disc, if we may so call it, of the nebulosity. In Prof. Bartlett's letter, mention is made only of the elastic force of the vaporous matter surrounding the comet and composing its envelope or tail. On this hypothesis, we have suggested the simplest and most natural method of completing the elliptic elements; viz. that of making the excess of the supposed elliptic, over the actual hyperbolic radius vector of the point observed, equal to a constant coefficient of the reciprocal of the square of the radius vector; and determining by means of the constancy of this value, the actual eccentricity corresponding to a period of 21 $7/8$ years, and a perihelion point of the center of gravity or head of the comet actually outside of the sun though nearly in contact with it. In fact this multiplier is not necessarily constant, nor necessarily a coefficient of the reciprocal of the square of the radius vector; still this hypothesis is the most simple and plausible that can be made, and is perhaps quite as complex as the nature of the question permits us to make.

As the subject of the physical organization of the head, tail, and nebulosity or envelope of comets, has been discussed by Sir

William Herschel, Olbers, Brandes and Bessel, with their character-istic genius and acumen, we deem it proper to consider the bearing of their opinions and researches on the present question.

Sir William Herschel* states that the kernel or head of the great comet of 1811, of about 1″ in diameter, could only be seen with high powers in his most powerful telescopes, and that with ordinary instruments he saw only the nebulosity or envelope; but that when the head or kernel was seen, it was seen within the envelope eccentric from the sun, or in other words the densest portion of the envelope or nebulosity was eccentric toward the sun. This is precisely the phenomenon observed by Messrs. Herrick and Bradley with reference to the disc of the nebulosity, though the kernel or head could not be seen. This also agrees with Bessel's remark that this comet seems to have thrown out nearly all its head in forming the nebulosity and tail.

We come next to Olbers's theory† of the formation of the envelope and tail of comets. This was promulgated in 1812, shortly after the appearance of the great comet of 1811. We do not recollect to have anywhere met with a translation of it. It is perhaps the only theory ever proposed, that explains all the phenomena observed respecting that comet. Olbers supposes that any particle composing the surface of the comet, or approaching from the frozen regions of space within a certain distance of the sun, is affected with a new repulsive force, resembling that which drives off substances from an excited prime conductor. These particles, thus polarized, he supposes to be thrown off from the head of the comet with a force proportioned to the mass of this head or nearly solid portion of the comet, and inversely as the square of the distance from the center of this head. The same particle in acquiring polarity with reference to the comet, acquires also polarity with reference to the mass of the sun, and is repelled by that mass, instead of being attracted by it with a force also varying inversely as the square of the distance from the sun. The origin of this polarity may be ascribed to the action of the sun's light or heat, or both. This particle, thus

---

* *Monatliche Correspondenz*, Vol. XXVIII. p. 459.
† Ibid Vol. XXV. p. 3. Prof. Norton's theory of comets' tails, read at this meeting, much resembles Olbers' theory.—S. C. W.

endued with one repulsive force, acting in the direction of the prolongation of the radius vector from the sun, and inversely as its square, and with another repulsive force, acting in the direction of the prolongation of its radius vector from the center of the comet, and inversely as its square, and with its original tangential velocity, at the time of parting with its actual cohesion with the comet, moves away in space in such a manner as not to return. Now, the geocentric position in the heavens with respect to the head of the comet, of any such particle, for any given elapsed time after it is thrown off from the comet's surface, may be readily computed, from the known tangential direction and velocity for any assumed values of the two repulsive forces of the comet and sun, for a unit of distance—say the Earth's mean distance from the sun. Olbers remarks, that the heliocentric orbit of such a particle must be a hyperbola; moreover, that the points in space where the two repulsive forces of the sun and comet make equilibrium, for any original direction of repulsion from the surface of the head, must have the portions of expelled matter more condensed than any portion of space between such points and the comet's head, thus forming, apparently, a hollow envelope or nebulosity, in the shape of a hyperboloid, having the head of the comet in its internal focus, and its apex toward the sun; the continuation of this hollow hyperboloid from the sun, beyond the parameter, so to speak, forms the tail of the comet. The shape of the hyperboloid, and, consequently, of the tail, depends upon the ratio of the repulsive forces of the sun and comet. If the one is determined by measure, the other can be computed from it. Olbers made measures of the shape of the visible section of this hyperboloid.

Brandes* gave this theory a thorough discussion, and finds analytically that the opinion of Olbers is true, that the envelope, if so caused, must be a hyperboloid; and then, from the observed dimensions of a section of this hyperboloid, as seen from the Earth, computes the ratio of the two repelling forces for that comet.

The theory of the formation of comets' tails seems to have made but little advances from 1812 till the return of Halley's comet in 1835; when the astronomers of Europe, with a full knowledge of

---

* *Monatliche Correspondenz*, Vol. XXVI, p 533.

Olber's theory, employed their powerful instruments to observe the tail and nebulosity of that comet with reference to it.* Struve remarked, that the densest point of the nebulosity was eccentric in that nebulosity, conformably to the observation of Messrs. Herrick and Bradley for the recent comet. But the most indefatigable observer was Bessel,† with the Königsberg heliometer. He detected a pendulous or vibratory motion of that portion of the nebulous matter, which was expelled from the hemisphere of the comet next the sun, resembling that of a magnet round the magnetic pole. He finds that with the addition of this pendulous motion of this streaming matter, or, in other words, of the apex of the hyperboloidal envelope of the comet in the plane of the orbit, and for fifty degrees or more on each side of the antipode of the comet's tangential direction, all the observed phenomena of the tail of Halley's comet may be explained. Bessel then proceeds to compute the repulsive force of the sun and comet on these expelled particles, from the observed shape of the tail. He also computes, for any point of the tail, how long the expelled particle has taken, since the date of its expulsion, to arrive at the point observed.

All these phenomena Bessel explains upon the supposition that there is no molecular attraction or repulsion between the particles thus expelled; but that their heliocentric motion is due to the tangential direction in which they are expelled from the comet, and to the two forces of repulsion of the sun and comet, and the orbital direction and velocity of the particle at the time of expulsion. The reason assigned by Bessel for not supposing a molecular connection between the particles of the nebulosity of the comet, is the fact, that the rays from a star seen by himself in the Königsberg heliometer, through this nebulosity, were not refracted by the medium of the nebulosity, but preserved, when seen through this medium, the same relative position with reference to any other fixed star, as when seen before entering the medium. Bessel argues that this could not have been a vapor, gas, or air, through which the star was seen, or it would have been refracted.

Having stated the principal points of the theories of Olbers and Bessel regarding the formation of comets' tails, it remains to

---

* Schumacher's *Astronomische Nachrichten,* Vol. XIII. p. 303.
† Ibid. Vol. XIII. p. 177. Also Schumacher's *Jahrbuch* for 1837, p. 142. Con. des Tems, 1839.

consider their bearing on the present question. It is obvious that the difference between the radius vector of the head, and of any particle of the nebulosity, one in the point aimed at for instance, would not be a simple multiplier of the reciprocal of the square of the radius vector, but would, on the contrary, include a term depending upon the reciprocal of the square of the distance of this particle from the center of the head. It may be shown, however, from the length and narrowness of the tail of the recent comet, that this latter term is nearly insensible, since the comet's expulsive force, compared with the sun's, must in this instance have been very small; for when acting in the normal to the radius vector, it was able to impress but a small normal velocity on the particles thrown off.

We conclude then, that whatever theory of the formation of comets' tails we adopt, if we suppose the particles of the nebulosity and tail to be material, and the densest portion of the former to be extended from the center of the comet's head toward the sun, the simplest method of deriving the true elements of the orbit of the head, from the computed elements of the orbit of the point observed; is that which we mentioned in our letter of the 25th of May last.

We subjoin the elements of the comet computed from our first normal places, by Professor Anderson, of Columbia College, New York, viz.—

Perihelion passage,	Feb. $27^{d}.579857$ m. t. Green.
Longitude of perihelion,	$279° \ 40' \ 35''.50$ ⎫
Long. of ascending node,	$15° \ 0' 56''.45$ ⎬ m. eq. March 30.
Inclination,	$34° \ 21' \ 6''$
Eccentricity,	1.0008560
Gaussian angle,	$2° \ 23' \ 0''.76$
Perihelion distance,	0.00415697
Mean daily motion retrograde,	$146''.50299$

Lieut. Gilliss read a communication, entitled, "History of the progress in establishing an Observatory at Washington City, Description of the Building erecting, and of the Instruments ordered for the Depôt of Charts and Instruments of the U. S. Navy, by Lieut. J. M. Gilliss, U. S. N."

In this paper, the author gives a succinct history of the depôt of charts and instruments under the charge of a bureau of the Navy Department, and describes the site which has been designated by the President of the United States for the construction of the buildings hereafter to be used as the depôt.

The site designated by him is on the north bank of the Potomac, in the southwestern part of the city of Washington; the plot containing a little more than 19 acres. Its north front is about 790, the east 1150, and west 700 feet long; and the spot selected for the building is at the intersection of the axes of two streets prolonged, one being in the meridian, the other in the prime vertical. This point is 100 feet above tide-water, has a north horizontal range one and a quarter, and south range eight miles, and is 105 yards from the east, 95 yards from the north, and 300 yards from the south enclosure, the last being the river. The hill is of gravel formation, with a surface stratum of dry brittle clay, through which water filters almost as freely as through gravel. An excavation eight feet deep has been made for the walls and piers of the entire building; and it is intended, for greater security, to leave a ditch ten feet wide by nine feet deep, surrounding the observatories.

The author then describes with minuteness the plans of the astronomical depôt and the magnetic observatory, and the manner in which it is proposed to execute them. He remarks that these were submitted by him to several of the most experienced European observers, and that they have met their approval. The paper concludes with a particular description of the great achromatic refractor, comet seeker, meridian transit, mural circle, transit for prime vertical, and the magnetic and meteorological instruments, which have been ordered in Europe for the use of the depôt. It is accompanied by drawings of the ground plan and elevation of the two buildings.

Mr. William C. Redfield of New York, read a communication "On Tides, and the Prevailing Currents of the Ocean and Atmosphere."

Mr. Redfield stated that the substance of this paper had been prepared in 1838, at the request of a gentleman attached to the

United States Exploring Expedition, together with a series of maps and charts on which were delineated the predominating systems of winds and currents in different oceans, derived from the log-books of voyagers and other sources. These maps had been lost, in the wreck of the *Peacock*, but the general remarks and statements which had accompanied them he now submitted to the Society.

In his remarks on tides, Mr. R. suggested the importance of good observations on the direction of the stream of flood tide in the *offings* of the islands and headlands visited by the expedition, especially in positions not exposed to the local influence of banks and shallows. He also suggested the inquiry whether the great oceanic tide-wave does not perform a great revolution, on each side of the equator, in each great ocean basin, moving westward in the intertropical latitudes and returning eastward in the higher latitudes; if, indeed, this question was not deemed already settled by the inquiries of Professor Whewell.

In treating of the great oceanic currents, Mr. R. noticed the course of revolution in the gulf-stream system of currents in the North Atlantic, and the counter-revolution of the northern offset from this system, which supplies in part the great polar current that sweeps along the coasts of Labrador and Newfoundland. It has been alleged by late writers, that this polar current passes from the shores of Newfoundland to those of Europe, in coincidence with the gulf stream; but Mr. R. maintained that its southerly course is continued along the American coast and into the ocean depths of the Atlantic, and that at its intersection with the course of the gulf stream it becomes mainly a subaqueous current. The proofs of this he found in the impelling effects of the deep polar current upon the icebergs which it drives into the gulf stream, or even across it, unless sooner dissolved by the warm superior current from the tropic, and in the reduced temperature of the more sluggish and diffused portion of the polar current, which continues its southwesterly course along the coast of the United States, to the westward of the gulf stream.

Systems of currents analogous to the foregoing, and proceeding from similar causes, Mr. R. described as prevailing in other oceans, but modified and controlled in their courses by the direction and contour of the continental coasts, and by the islands and extensive coral reefs of the Pacific. Thus, his inquiries had shown a warm current in the western Pacific, corresponding to the gulf stream of the Atlantic, and which had been incidentally noticed by the officers of Cook's third expedition, as running on some occasions with a velocity of five miles an hour.

After some allusions to the geological effects of the permanent systems of ocean currents, Mr. Redfield proceeded to notice the predominating currents of the atmosphere, as exhibited in the unceasing and mainly horizontal movements of the air, in circuits or systems of revolution and compensation, in the several oceanic basins of the Earth's surface, on both sides of the equator. Of these systems of circulation and revolution, which bear a general resemblance to those of the ocean currents, the portions which move westwardly in the lower latitudes constitute the so called trade winds; while the opposite winds, or counterparts of the several circuits, are found in the more gyratory and irregular winds of the temperate zones, which maintain a general movement to the eastward, that fully compensates for the westwardly flow of the trade winds. The immediate but often unseen connection of these opposite winds in continuous parabolic or elliptic circuits, of varying forms and extent, is more conspicuous in some regions than in others, and is most clearly developed near the extreme eastern and western borders of an ocean basin; as in the region which includes the island of Madeira and the Canaries, in which northwardly winds are found to prevail; while near the western margin of the same ocean, in like latitudes, southerly winds are most often met with, except in storms or at particular seasons.

The major axes of revolution, in these great natural circuits of wind, are generally found in those extensive belts or regions of extratropical calms and light winds, which are found in all the great oceans, and which, in the North Atlantic, are known among mariners as the "horse latitudes." A more extensive belt of calms and light winds, as is well known, generally separates the trade winds and revolving systems of the two hemispheres, in the regions near the equator.

The generally horizontal winds of our globe, as thus developed in the great systems of circulation in its atmosphere, Mr. Redfield had referred on some former occasions to the law of gravitation, as connected with the rotary and orbital movements of the different geographical parallels and meridians of the Earth's crust; the nearest attainable equilibrium of the fluids which envelop the planet being that of motion, not of rest. But a different theory of winds has been generally adopted, founded on the alleged effects of rarefaction in the equatorial regions. Without discussing these theories, and while admitting heat as a cause and modifier of winds to some considerable extent, Mr. R. now alleges the following as valid and insuperable objections to the common theory of the trade winds, viz.

1st. The specific difference of temperature in the inter-tropical winds, as compared with equal zones of extra-tropical winds, is inadequate and disproportioned to the dynamical effects exhibited in these winds.

2d. The ascent of the body of trade wind to the upper atmosphere, in the equatorial latitudes, has never been shown, by observation, and may well be denied.

3d. The perpetual snow line of the Andes is found to be near 1000 feet higher in 16° to 18° south latitude than at the equator or on the parallels of the equatorial calms.

4th. The semi-annual change in the locality of the zone of greatest heat is not productive of any like degree of change in the locality of the trade winds.

5th. The course of the winds in extensive portions of the torrid zone appears wholly irreconcilable with the received theory.

6th. In our American summers, the hottest winds are often found moving horizontally on the Earth's surface, for several successive days, but not toward the equator, nor rising from the surface, as the theory requires.

Further proofs of the horizontal course of revolution in the trades and the general winds, in opposition to the calorific theory, was found in the known progress and routes of those extensive portions of the lower atmosphere which comprise the great storms, as shown by the inquiries of himself, Col. Reid, and others. These extensive portions of atmosphere cannot be supposed to possess any self-moving power or tendency other than gravitation, but must move in accordance with the predominating physical impulse to which the lower atmosphere is subject in the regions through which the storm may pass.

The *rotation* of these storms, also, in a determinate direction, proves the generally horizontal course of the winds on the Earth's surface, and the dynamical influences of the Earth's rotation on the atmosphere; for it can be shown by an experiment which Mr. R. pointed out, that a tendency to gyrate from right to left in the northern hemisphere, and in the counter direction in the southern, is necessarily imparted to the surface winds, which move from the equator toward the temperate latitudes, in the great circuits of revolution which he had ascribed to the trades and to the general winds.

Dr. Horner, Professor of Anatomy in the University of Pennsylvania, presented a summary view of the existing application of the microscope to human anatomy and to animal organization generally, and considered many of their important problems as solved by its recent improvements.

He alluded to the observations of the earlier microscopical anatomists, and the difficulties they experienced from the want of an adequate provision against spherical and chromatic aberration. These difficulties had at length been surmounted, so as to afford greater uniformity between different observers, and thereby to produce greater confidence in the results. Many points may, indeed, be considered as definitively settled. He spoke highly of the compound microscope of Ploss, but gave the award of decided superiority to that of Powell and Lealand, of London, which had been introduced to the notice of the citizens of Philadelphia, by Dr. Ch. Fr. Beck.

The points of microscopic anatomy, to which he referred specially in illustration of his views, were the settlement of the questions on the diameter and structure of the blood discs or corpuscles, on the chyle and lymph corpuscles, on the fat vesicles, and on the texture of the cuticle and epithelium.

He also alluded to the discovery of Schwann, that the rudimentary state of every texture and organ is that of cell, and that from its modified evolution every thing else was developed; a principle in histogeny more prolific in consequences than any other which had ever been asserted.

He likewise alluded to the cellular origin of the colors of different parts of the human body, that they were all, including that of the skin and choroid coat of the eye, generated in pigment cells.

He introduced, further, a notice of the microscopic organization of the nails, teeth, muscular tissue, and mentioned the point, now decidedly ascertained, of the nervous fibrillæ running their course from origin to termination without anastomosis, and of their being in the condition of tubes.

In this limited notice of the innumerable objects of microscopical observation, he considered himself as doing very inadequate justice to the labors of the gentlemen who had enlarged so considerably the boundaries of human knowledge; but the technical character of the subjects compelled him to make only a brief exposition of what had been done: his chief notice being to show, that while mathematics, natural philosophy, natural history, and chemistry, are pushing forward their confines, anatomy has not been idle.

Professor Bache presented the results of two year's' observations of the magnetic elements, and of the temperature, pressure, and moisture of the atmosphere, at the Magnetic Observatory at the Girard College.

The Observatory was opened in May 1840, and the results are for two years from that time. Prof. Bache, after alluding generally to the different results obtained from the magnetic observations, called particular attention to the averages of the hourly changes. The observed elements were the declination, and the horizontal and vertical forces. The changes of each of these were represented by curves in the usual way, the differences of the abscissæ representing the times elapsed, and the differences of the ordinates the change of element; increasing ordinates corresponding to increase of declination or of force. The curves, even with the limited number of observations upon which they are founded, are remarkable for their regularity.

The declination shows distinctly two maxima and two minima during the twenty-four hours. The chief maximum is about 1 P. M., and the subordinate maximum about 1 A. M.; the latter, though marked, does not reach the mean line of the twenty-four hours. The minima are at about 8 A. M. and 9 P. M. The mid-day maximum was about 8′ above the preceding, and 6′ above the succeeding minimum, and the midnight maximum 1′ above the next preceding, and 3′ above the next succeeding minimum. The line of mean declination crosses the curve about 10 A. M. and 6 P. M. The force observations were, Prof. Bache remarked, corrected approximately for temperature; and he made some statements to show how difficult it was to obtain a correction for temperature, especially in large magnetic bars. He expressed himself not satisfied with the correction yet obtained, although it resulted from a very large number of comparisons, and appeared to represent reasonably well the average correction.

The curves representing the horizontal and vertical force observations were traced separately. The horizontal force shows, like the declination, two tides in the twenty-four hours; the vertical force but one. The maxima of the horizontal force are at about 6 A. M. and 2 P. M., and the minima about noon and 9 P. M. The maximum of the vertical force is at about 1 P. M., and the minimum about 9 P. M. Prof. Bache remarked upon the changes of total force and dip, as indicated by those of the horizontal and vertical force. The line of mean horizontal force crosses the curve four times, at about 8 A. M., 2 P. M., 5 P. M., and $11^1/_2$ P. M. The line of mean vertical force crosses the curve twice, at about 6 A. M. and 6 P. M. The range of horizontal force, or difference between the greatest maximum of the twenty-four hours and the lesser minimum, is about .001,092, expressed in parts of the entire horizontal force; and the same difference for the vertical force is about .001,024. These observations had been made the basis of a series to determine the times of maxima of the several elements more nearly, by observations taken at short intervals, at the periods of the day, including these times.

Prof. Bache gave reasons why he considered the phenomena thus presented to be inconsistent with the idea, that they are produced by the heating effects of the sun, acting either directly or indirectly.

The averages of meteorological observations were of the thermometer and barometer; the force of vapor also was given as deduced by Prof. Apjohn's formula and the tables in the instructions of the Committee of Physics of the Royal Society, from the observations of the wet bulb hygrometer. The hourly means for the two years, and the means for each month were represented by curves. The curve of hourly change of temperature was of the parabolic form frequently before obtained, with irregularities depending upon the number of observations. The periods of maxima and minima were about 3 P. M. and 5 A. M. The curve representing the force of vapor resembles, generally, that of temperature, though it is by no means as regular; the maximum being, however, about 6 P. M., and the minimum between 4 and 5 A. M. The curve of pressure shows the two atmospheric tides, the maxima at about 9 A. M. and midnight, and the minima at 2 A. M. and 5 P. M. The mean barometric fluctuation in twenty-four hours is about .06 of an inch. The similarity of the curves of monthly means of temperature and force of vapor is even greater than that of the corresponding curves of hourly means. The maximum temperature is reached about the middle of July, and the maximum force of vapor later in the same month; the minima in December and January. The curves of pressure prove, that an element besides the temperature and force of vapor must enter into its composition. Comparisons of the numbers for the force of wind had not yet been made, though the data had been obtained.

Prof. Bache observed, in conclusion, that as the Observatory had been re-opened by aid obtained from the Secretary of War, similar observations would by degrees be accumulated, and by an increased number of results render the means more trustworthy than those now presented.

At the close of his remarks, Prof. B. invited the members and correspondents of the Society and the gentlemen attending its meetings, to inspect the Observatory at the Girard College after the adjournment in the afternoon.

Mr. Henry D. Gilpin, pursuant to appointment under Chapter I., Section 13, of the Laws, presented a Biographical Notice of the Hon. Edward Livingston, late a member of the Society.

Mr. Gilpin introduced the designated subject of his paper by claiming the honors of distinguished membership in our Society for all who, whatever their pursuits of life, have illustrated our country by their intelligence, or happily guided it by their wisdom. Without detracting from the merit and the praise of those, who in the pursuit of natural or exact science and the elucidation of moral truths, more usually win, in the general, estimation of the world, the cherished titles of philosophy; no one, he said, does philosophy claim more justly and truly as her son, than him, who in the active engagements of a public career, where he is stimulated by ambition, and occupied by objects that are supposed most strongly to absorb the feelings, if not to warp the judgment and the taste, yet blends with all his actions the love of science and the extension of truth, and applies that wisdom which springs only from knowledge and truth, to the affairs he is called upon to engage in or direct. Exemplifying this observation by the names of Aristotle, Cicero and Bacon, and by those of Jefferson and Franklin, Mr. Gilpin asserts for that of Mr. Livingston a place in the same category with theirs. He points to the dignity of his intellectual labors, the light which they reflect of philosophic thought and large experience, their eminent practical value, and the consistent anxiety which they always evince for the advancement of human happiness and virtue. With a mind, he says, clear, penetrating, and sagacious; with an industry that left unfinished no duty that he undertook, there was blended from his earliest youth a serenity of temper, simplicity, and cheerfulness of manners and active benevolence, a dear strong sense of right, a desire to promote in all things the good of others, and a willingness to forego his own interest and inclinations; so that in all the relations of an active and varied life, he filled his part, not more for his own enduring reputation, than for the benefit of those he served.

Born but a short time before the commencement of the revolution, the youthful years of Mr. Livingston were impregnated with the lessons best taught to an observing mind by the incidents that occurred around him. A brother of Robert R. Livingston, one of the Committee who draughted the Declaration of Independence; a brother-in-law of Montgomery, who sealed with his blood the manifesto of patriotic resistance; filled with an insatiable love of study, by which he had mastered the stores of ancient and modern learning, and acquired a knowledge, far from inconsiderable, of many of the branches of abstract and natural science; he came into life just at the period when the institutions of his country assumed their settled form, imbued with the true spirit in which they were founded, animated with the desire to maintain them in purity and vigor, and possessing the talents and information which would enable him well to perform his part, in whatever situation he might be thrown, as a private or a public man. Having adopted the legal profession, and established himself in the City of New York, he had gained before he reached the age of thirty, a high reputation for the extent of his acquirements as a jurist, and ability as an advocate. He already began to apply to his professional investigations the principles which be had accumulated in the wide range of his legal studies, and in some degree introduced into a practice, necessarily founded upon and nearly confined to the English law, those illustrations which he had derived from the jurists of antiquity and of continental Europe, and which at a subsequent period were so conspicuously and advantageously exhibited in his public and professional labors. From these occupations he was in some degree withdrawn for several years by his election to Congress, as a representative from the City of New York. That event took place in the year 1794. He was twice reelected; and during the six years that he remained in Congress, he maintained a position equally distinguished by the ability that marked his views on all public questions, and the enlightened and candid spirit which he evinced in the discussions of a period, when those differences were first developed, that presently assumed a character more ardent and limits more distinctly defined. United in political opinions with Madison, Gallatin, Giles, and Macon, he bore a conspicuous share in the debates on the public measures which they advocated or opposed; and came at once to be considered as a leading member of the party to which he attached himself. Independently, however, of subjects more peculiarly political, he was early the advocate of various measures indicative of a wise and philanthropic spirit; and among these, it is especially to his exertions that we owe the first endeavors to reform the criminal code of the United States, to protect or relieve American seamen left by accident or misfortune on foreign shores, and to promote the gradual increase of a navy sufficient to protect American commerce in remote seas. Though sincerely attached from a strong conviction to the particular political opinions which he advocated, he yet maintained them with characteristic liberality and deference to those from whom he differed, and no diversity of views of general policy could induce him to withhold his cordial support to such of their measures as he deemed calculated to sustain or protect his country's honor or rights.

Soon after withdrawing from Congress, he was elected Mayor of the City of New York, an office, whose organization as then constituted required the exercise of important judicial as well as executive functions. He was also selected by Mr. Jefferson, when he became President, to fill the post, for which his legal acquirements eminently fitted him, of Attorney of the United States for the State of New York. During the period of his mayoralty, the city was afflicted by a desolating pestilence; during which his personal exertions and benevolence were fearlessly displayed at the risk, and almost with the loss of his own life. The Common Council, on his subsequent retirement from office, adverted, in an address unanimously adopted, to his admirable conduct in that trying emergency and in the whole performance of his public duties, and expressed to him the gratitude of the community in glowing terms.

When Louisiana had been gained for the United States by the diplomatic labors of his brother the Chancellor, Mr. Livingston resolved to remove there, and to connect his renewed professional career with the rising institutions of the new community. The enlarged nature of his earlier legal studies, enabling him at once to grasp the questions which arose out of the provisions of the civil law, as well as that of France and Spain, introduced there at different periods of colonial authority; his thorough knowledge of the jurisprudence then generally prevailing in the United States, which would necessarily come to be incorporated to some extent with that of a territory now a part of them; and above all his habit and power of careful discrimination of legal principles, looking to them according to their intrinsic excellence and fitness, neither in a spirit of unnecessary innovation nor an unwise adherence to mere precedent or usage; these qualities not only placed him at once, by general consent, at the head of his profession in Louisiana, but they enabled him to exercise a more than common influence in establishing a system of jurisprudence there, which in all respects may bear an advantageous comparison with that of any other of the states, and may claim over that of many of them a decided superiority.

Immediately on his arrival in Louisiana, Mr. Livingston perceived the necessity of prompt attention to this subject. The inhabitants had grown up and were living under a system of laws, which were henceforth to be administered by judges, some of whom were ignorant of the languages in which they had been promulgated, and most of whom had been accustomed to judicial forms altogether different. On the other hand, the institutions of a free people were to supplant those of a monarchical and colonial government, and the individual citizen was to be called on constantly to perform a part personally active and efficient. It became therefore at once essential that, even without a change in the body of the laws, a mode of procedure fitted to the circumstances should be established without delay. The legislature wisely resolved to commit this duty to the judgment and knowledge of jurists, in whom the people might safely repose unlimited confidence. Mr. Livingston was selected to perform it, and with him was united a personal and professional friend of learning and ability, also while he lived a member of our Society, Mr. James Brown, afterwards a Senator in Congress from Louisiana, and Minister Plenipotentiary to France. While they discarded the fictions and technicalities of the English law, they avoided the prolixity so usual in the Spanish, and not infrequent in the French code. Their system is simple and intelligible, well calculated to prevent unnecessary expense and delay. It was adopted by the legislative council; it was introduced with the general approbation of the community; and for a series of years it has stood, with slight alteration, the test of trial and experience.

A more important task remained—the complete revision of the body of civil and criminal law; and its reduction into systematic codes. This was naturally and properly postponed until, by their admission into the Union as a sovereign state, the people could themselves act upon a measure so important to their feelings and welfare. After this event, with a just estimate of the wisdom and ability of Mr. Livingston, he was selected for the task by the Legislature: no difference of opinion upon the political topics of the day withdrew their confidence from one, who had identified his fame with the jurisprudence of Louisiana, as he had devoted his talents to their service. In the preparation of the civil code, Mr. Dubigny and Mr. Moreau were united with Mr. Livingston. The task proved to be one of great labor: the existing laws, which were familiar to the people and therefore not without necessity to be abolished, consisted of provisions at once complicated and discordant; the fragments of the Spanish ordinances frequently remained; the French law previous to the revolution had not been altogether superseded by the code of Napoleon; and, with American judges, and the influx of American citizens, many of the provisions of the English common law had obtained a place. The arduous exertions of three years were required to reduce this mass into an intelligible shape. In all its parts it received the cooperation of Mr. Livingston; and some of them, especially the title of "obligations," were exclusively his own. It met with a reception from the Legislature and people of the state far more favorable than could have been anticipated for such a measure; and with the exception of the commercial code, to some provisions of which objection was made, it was promptly adopted and still continues, with few alterations in its general principles, to be the permanent law of the state.

By an act of the Legislature the preparation of a system of criminal jurisprudence was confided to Mr. Livingston alone. He deeply felt the responsibility he assumed in undertaking such a trust; he knew that he would have to encounter strong prejudices, to oppose long-settled opinions, to exercise a vigilance of forecast and distinctness of enactment as to the objects, for which the state of society present and future required him to provide. Two years after his appointment, he presented to the Legislature a preliminary report, exhibiting the progress he had made, explaining the plan on which he proposed to execute the work, and giving some detailed parts as specimens of it. These were unanimously approved, and he was requested by a vote to complete his labors. He accordingly proceeded with it. His best faculties, to use his own language, were faithfully and laboriously employed, under the direction of a religious desire to perform the duty entrusted to him in a manner that might realize in some degree the views of his fellow citizens, for whose benefit it was designed. By assiduous exertion he completed the entire work in two years more; but it was scarcely finished when all his labors were destroyed by an accident, that fortunately, in its final result, only produced a remarkable instance of his equanimity and perseverance. Having received authority from the Legislature to submit it to them, when completed, for greater convenience in a printed form, he had caused a fair copy of the whole work to be written for the use of the printer. The evening before it was to be delivered to him, he occupied himself till a late hour in comparing this copy with the original draught. He left them together when he went to bed, consoling himself with the pleasing thought that he had thus completed the labors of four years. Not long afterward he was awakened by the cry of fire; he hastened to the room where his papers had been left, but not a vestige of either copy remained. They were totally consumed. Though stunned at first by the event, his industry and equanimity soon came to his aid; before the next day closed, he had recommenced his task; the Legislature at their following session extended the period for its performance; and in two years more, he presented to them his complete "System of Penal Law," in the shape in which we now see it. Prefixed to the system was a series of reports, reviewing in a masterly manner the whole science of penal jurisprudence; pointing out the objects to be sought for, the errors to be combated, and the modes in which these could be done with most benefit to the criminal himself and to the society whose laws he had violated. The system has not, it is believed, been yet finally acted upon, in its extended form, by the Legislature of Louisiana; but it does not on this account claim less justly the admiration of the philanthropist and jurist. It is a work worthy of the deep consideration of all communities. The beauty of its arrangement, the wisdom of its provisions, and the simplicity of its forms, have never been surpassed, probably never equaled in any similar work; and it is not without entire justice that this admirable production has contributed, perhaps more than any other of his labors, to secure to Mr. Livingston that eminent place which he holds among those who are regarded not merely as distinguished jurists but as public benefactors.

It was by these acts, during an uninterrupted residence of many years, that Mr. Livingston identified himself with the state of which he became a citizen. His name will ever be cherished with grateful affection and respect in Louisiana. Nor was it by these acts alone. His eminent standing in his profession and in society, the active interest which he took in all the institutions of the state, and his services in the Legislature, of which he was occasionally a member, all united to make him not only an influential citizen, but one who was able in innumerable ways to contribute largely to the benefit of the community. His patient industry, his amenity of temper, the generosity of his disposition, made this at once easy and agreeable; and when, in the circumstances of the times, acts of more serious devotion to public duty were required, he was found amongst the foremost, ready and zealous to discharge them. The invasion of the British at the close of the war roused the patriotic spirit, as it required the prompt devotion of the inhabitants. With but few regular troops, and almost entirely unprepared for such a conflict, they were obliged hastily to form themselves into an army to repel the invaders. Mr. Livingston was among the foremost to do so. Instantly leaving his professional duties and all private occupations, he presented himself to General Jackson as soon as he arrived to take the command in Louisiana, and offered to place himself in any position where the General might regard his services as useful. He was selected as his aid-de-camp. He was by his side constantly throughout the period of hostilities, enjoyed his confidence in a marked degree, and at the close of the war received from him many evidences of that regard which was afterwards, and in another station, yet more signally displayed.

After an uninterrupted residence in Louisiana for twenty years, in which he had withdrawn from political pursuits, and devoted himself to his profession and those congenial studies and labors that have been adverted to, Mr. Livingston determined to retire from the bar, and to revisit in New York the scenes of his earlier life, and the connections from whom he had been so long separated. This determination was the signal for a new mark of confidence from his adopted state. He was elected as a representative in Congress from Louisiana, an event that was followed in a few years by his choice as a senator. After his election, an enthusiastic address was presented to him by the City Council of New Orleans, in which they reviewed his various public services from the moment of his arrival in Louisiana, spoke of them in warm terms of approbation and gratitude, and expressed their confidence that his continuance in the national councils would be a sure guarantee of further exertions for their welfare and prosperity.

Mr. Livingston continued in Congress from 1823 to 1831. His advanced age prevented the same energetic participation in the public business, which had there formerly distinguished him, but he nevertheless, originated several important measures, and not unfrequently engaged in debate. The speeches that have been preserved exhibit that clearness of perception and language, that various but unostentatious learning, that simplicity, dignity, and patriotism, which were characteristic of him. His views of public questions were expressed with firmness, but without asperity: he discussed, in a masterly manner, all those topics connected with a true construction of the Constitution, and the extent and limitations of power assigned to the members of the confederacy and to the different departments of the government, which grew out of the controversy in South Carolina: the reputation which he had acquired in Congress so long before, and that which had been added to it by his eminent labors in a different sphere, suffered no diminution, but gained additional luster by his return to a legislative career.

In the spring of 1831, the Department of State became vacant by the resignation of Mr. Van Buren. Mr. Livingston, who had retired a few months before, at the close of the session of Congress, to an estate that he possessed on the Hudson river in the neighborhood of his birthplace, was summoned by General Jackson to fill that elevated post. Totally unprepared for such an event, he hesitated for some time to accept it: with the modesty and simplicity that marked his character, he distrusted his abilities adequately to discharge its duties; and it was not without difficulty that the President obtained the services of one, whose devotion to his country he had himself witnessed in far different scenes, and whose talents and virtues had received the approbation of his countrymen so often, and in so many ways. Eminent as have been the men who have filled the post of Secretary of State, few have displayed the same fitness and ability to discharge its duties. His negotiations with foreign nations were very successful; and the documents connected with them, so far as they have been published, exhibit profound political wisdom, and an enlightened spirit. The treaties that he formed are not more beneficial in their commercial stipulations, than they are made consonant, in their international provisions, with the feelings and improvement of the age. The missions that he originated or promoted have opened new and important fields to American enterprise. The counsels of which, as the chief member of the administration, he was the advocate or adviser, were founded on views of the constitution carefully considered and ably vindicated.

The duties of such a place were, however, more arduous than Mr. Livingston, at his advanced age, was willing to continue long to discharge; and on the reelection of General Jackson in 1833, he retired from his cabinet. At that time the negotiations with France, arising out of the treaty of 1830, which granted an indemnity to the United States for injuries done to American commerce during the wars of Napoleon, were in a state of great complexity. This was increased by the excitement which party contests in the French legislature gave to the subject; and it was evident that the position of affairs demanded such a course on the part of the United States as should protect its honor and maintain its rights, without allowing any thing, not required by these just objects, to interfere with or endanger that ancient friendship between the two nations which had its origin in the struggles of the revolution. For such a service, no man in the United States was more eminently fitted than Mr. Livingston. The distinguished public office from which he had just retired, the ability and consistency which had marked his career as a statesman, his sound views in regard to the institutions and policy of his country, made him a representative of American feelings, opinions, and determination, in whom his fellow citizens had a perfect confidence. The known moderation of his character, his reputation as a jurist, especially on international questions, his long residence in Louisiana, whose inhabitants were connected with France by so many associations, his knowledge, which was more than commonly profound, of the language, literature, and history of that country, seemed to assure for him the most friendly reception there; and, as if to add to these circumstances of peculiar fitness for such a post, he had not long before been elected a member of the Institute of France: so that he was already enrolled among a body of distinguished Frenchmen, and connected with them by those ties which spring from mutual labors in the paths of science and of philanthropy, and in the search of wisdom and truth. He was accordingly selected, in the summer of 1833, by President Jackson, to fill the post of Minister Plenipotentiary to France. He accepted the appointment, embarked shortly after in the Delaware ship of the line, and arrived at Cherbourg in the month of September. He remained abroad until April 1835, when he returned to the United States. Although, at the time of his leaving France, the differences between the two countries had not been finally adjusted, and, his departure was a step taken in consequence of what he deemed due to the honor of his own country, yet it was shortly afterward followed by an acquiescence on the part of the French government in the course which, under the instructions of President Jackson, he had firmly but temperately urged. His whole conduct, in circumstances that demanded at every step the exercise of an able judgment and an enlightened patriotism, served well to terminate his career as a public servant; and the official documents in which it is exhibited and vindicated, must ever be regarded as among the most excellent of his own state papers, and will deservedly hold a conspicuous place in the history of our intercourse with foreign nations.

Mr. Livingston did not long survive his return to his native country. He immediately resumed his residence at his estate on the Hudson river, among his numerous family connections, and the rest of his life was spent in scenes rendered attractive to him at once by their own natural beauty, and by the associations of his earlier years. He devoted himself with the greatest enjoyment to the pursuits of the country. His farm and his garden, with that social intercourse in which he always loved to indulge, afforded him constant employment; and it was in the midst of such occupations that his life was terminated, by a sudden illness, in the spring of 1836. He had just reached the age of seventy-two.

The private life of Mr. Livingston was a daily exhibition of domestic and social qualities which secure affection and diffuse happiness; his temper was serene and his disposition cheerful; his heart was keenly alive to all the impulses of affection and of friendship; he could bear misfortune with equanimity, but to the close of life readily participated in the cheerful amusements of society; devotedly fond of study, and having untiring industry and a retentive memory, his mind was richly stored with all the knowledge that literature could impart; fond of scientific investigations, so far as his many engagements permitted him to pursue them, he readily gave his aid to those who engaged in them, actively benevolent, he was unceasing in his endeavors to promote every plan which he deemed conducive to the welfare or improvement of men. In his profession he was eminently distinguished; as an advocate and a lawyer, he stood by general consent in the highest rank; and his labors in those kindred branches of study and reflection, which were required in the preparation of the systems of civil and criminal law which he framed, gave him a reputation, and secured to him honors and distinction in his own and other countries, not surpassed by any of the jurists of his times. Among the statesmen of America his place was no less eminent; his public speeches present in every instance striking views of the questions he discussed, and although the stations of trust to which he was elevated place his official labors in comparison with some of the most illustrious of his countrymen, this has only served to display more clearly their intrinsic merit, and to secure for them an equal approbation.

SPECIAL MEETING.

Fourth Session, 27th May, half past 7 o'clock, P. M.

DR. BACHE, Vice-President, in the Chair.

A letter was received from Dr. John Locke, of Cincinnati, Ohio, containing a brief notice of the method which he had adopted for replacing the cross hairs in the telescope of a transit instrument.

Dr. Locke remarks, that though this method may not be new to instrument makers, the description may be of service to amateurs who are at a distance from such aid. Dr. Locke used the threads taken from a spider's cocoon. The points to be accomplished were to stretch the line to a perfect tension in a manageable way, to remove from it the flexuosities of the cocoon, and to adjust the lines parallel to, and equidistant from, each other. The first condition was answered by fastening the thread with wax between the ends of an elastic wire, bent in the form of the letter "U," and the ends of which were made to approach each other, slightly, by a wooden clamp. The line was straightened by holding it, while under tension, above a vessel of boiling water. Each line in turn, while still stretched, was put in place by the lines engraved upon the diaphragm, verified by a micrometer microscope and fastened by varnish. In the final placing of the lines a small wire, sliding by means of a fine screw, was used to change slightly the place of the several fibers. Dr. Locke states that he has had a diaphragm made with adjustible fibers, so as to avoid any error in the intervals. The fibers are fixed at one end only, and are brought into position by resting in their course across the plate upon oval pins, which can be turned by square heads formed upon them.

A communication "On the Launch of the Three-deck Ship, the *Pennsylvania,* in 1837," was presented by Mr. John Lenthall, Naval Constructor U. S. N., and was read by Mr. J. C. Cresson.

Mr. L. prefaced the description of the launch with some remarks on the difficulties arising from the decayed state of the wharf at the Navy Yard, which rendered it necessary to rebuild the foundation of the ship after considerable progress had been made in her construction.

The building of the ship was commenced in the latter part of the year 1822, and she was launched in July 1837, leaving Philadelphia in November of the same year.

Her dimensions are,—

Length at the upper deck,..............................223 feet 7 inches.
Extreme breadth,.. 58 " 2 "
Height from lower side of the keel to the top
    of the rail,.. 54 " 8 "

Her armament will consist of 130 guns:

Lower gun deck,	4 8	inch Paixhan guns,	and 28 light 32 pounders.
Middle "	4 8	" "	and 30 "
Upper "	8 8	" "	and 22 "
Spar "	2 32	pound	and 26 32 lb. carronades.

The total cost of this ship when she left Philadelphia, exclusive of the armament, was $687,026, of which about $535,000 was expended on the hull before she was launched.

The weight of the hull when she was launched was estimated at 2696 tons, of 2240 lbs.; the weight upon the bilge-ways being 2720 tons, including extra weights on board.

The inclination of the ways and of the keel, was 3° 31', making the force down the plane 166 46/100 tons, and the pressure on the plane 2714 33/100 tons: allowing the friction to be 0.05 or 1/20 the pressure, there was a tendency of 30 73/100 tons to cause the ship to descend the plane at the commencement of the motion.

The paper then proceeds to give a detailed description of the preparation of the launching and bilge-ways, and the methods adopted to give entire security to the immense structure when it should be put in motion; and gives the particulars of the dimension of the timbers, modes of fastening, &c., of much value and interest to practical men.

The immediate dispositions for the launch were begun at five o'clock on the morning of the 18th of July; and shortly after two in the afternoon, all the shores and blocks having been removed, the extended plank of the bilge-way which formed the last attachment of the vessel was sawed off. She did not move, however, in consequence of the too great rigidity of the mixture with which the ways were greased, until an iron wedge had been driven into an opening made in the bilge-way plank by sawing out a small section, and an impetus was given by levers and suspended rams, which had been arranged in anticipation. Yielding readily to the force so applied, she entered the water at twenty minutes past two, without the slightest misadventure. Her draft of water was found to be eighteen feet seven inches aft, and fourteen feet ten inches forward.

Mr. Kane read a letter from the Rev. Professor Alonzo Potter, of Union College, Schenectady, N. Y., containing a brief reference to the career of Chancellor Livingston, and his useful labors in the cause of science and the arts. Professor Potter's letter was accompanied by an autograph letter of Count Rumford to Chancellor L., dated Auteuil, near Paris, 8th December, 1811, and a letter from Chancellor Livingston to Dr. De Witt, Secretary to the Society for Useful Arts, &c., dated Clermont, 26th November, 1806.

The following extracts are made from these:—

Count Rumford to Chancellor Livingston, 8th Dec. 1811.

I rejoiced to hear of the success which your steam-boats have had, and was glad to hear from yourself that the invention is likely to become extensively useful.

You have but one thing left to do; and that is, to establish a steam-wagon to carry passengers by land. I have not the least doubt of the practicability of the scheme, and do much expect to live long enough to see it executed. The machinery must be worked with strong steam, and no attempts must be made to condense it. As the resistance, on a flat road, will be very nearly the same with a swift motion as with a slow one, you might travel in this way with astonishing celerity, taking care to moderate the motion in turning corners. I imagine, that by making the wheels very high, and of cast iron, they might be made to act as flys, to equalize the distribution of the force. A fly of some kind or other would, I fancy, be indispensably necessary. Broad wheels and good roads would certainly be necessary.

Apropos of broad wheels: I send you enclosed a paper, in French, on that subject. I am quite sure they will come into fashion sooner or later, for all sorts of carriages. I use no other, and many persons are following my example in Paris and other places.

I send you another paper on lamps, which has made a more sudden revolution in this country. The portable lamps are now to be seen in several shops in Paris, and the lampe-à-colonne, for dining-tables and drawing-rooms, is in very general use. I wish I knew how to send you one of each of them. You will hear of another lamp soon, destined for light-houses, which gives just as much light as you please.

I have one, which with four flat wicks, each an inch and a half broad, placed at the distance of an inch from each other, with air coming up between them, gives as much light as fifty-two wax candles all burning together. By your consul, Mr. Russel, I lately sent an account of this new invention to the Royal Society, and from their transactions you will probably learn more of the matter in a few months.

I am just now employed in making a course of experiments on the quantities of heat produced in the complete combustion of various kinds of wood, and other inflammable substances.

Chancellor Livingston to Dr. De Witt, 26th *Nov.* 1806.

I forgot, when I had the pleasure of seeing you, to mention an invention which might, if perfected, be rendered of very general utility. While at Paris, I ordered a carriage for the purpose of trying it, but I was called away by the sailing of my ship before I could execute it. The object of it was to contrive some better springs for carriages than those now in use. Every body knows the utility of springs in saving the traveler from fatigue, and the carriage from being jolted to pieces in rough roads. But it is not so generally known, that they enable a horse to go through his work with much less fatigue; could they therefore be adapted to farming carts, they would be found extremely useful. The springs of carriages now in use are made either of wood or iron. The first is too weak or too clumsy; the last is not only expensive, but heavy and liable to rust, and above all to snap in very cold weather. Springs of either of these materials have one common and great inconvenience, that of not being able to adjust themselves to the different weights that are placed upon them. If they are so stiff as to bear a heavy burden, they have no elasticity under a light one; or if they spring under a small pressure, they break under a heavy one. This circumstance greatly limits their utility. For wood and iron I would therefore substitute the lightest, the cheapest, and the most elastic of all substances—air. This can never break, and its spring will always be proportioned to the weight that it acts upon. Place a carriage box upon the pistons of four brass tubes, each containing twenty inches of air. If these were four inches deep, it would require 295 lbs. to sink the pistons two inches, and four times that weight, or 1180 lbs., to sink them three inches; and upwards of a ton weight to bring them half an inch lower: in every case the spring would continue to act with a force proportionate to the pressure. If a greater motion in the spring is required, let the tubes be deeper. If eight inches deep, the motion, under equal pressure, will be the double of those I have mentioned. There are various ways in which these springs may be adapted to carriages. Of these, perhaps, the cheapest and the best would be two planks, united by leather dressed in oil, and covered with elastic gum, so as to be perfectly air-tight. For a chair, four bladders soaked in oil, and covered with strong leather in the way of a foot-ball, would make a cheap and excellent spring. The leathers should be put on before the bladders are blown up, and be somewhat smaller than the bladders, so as to press them strongly in every part; this would keep them from breaking, or losing any air when strained. These balls should be confined in boxes that fit to their lower diameter, and over these the thorough braces that hold the chair should pass, and be fastened to the bars before and behind the chair. This would not only render such a carriage much lighter than those now in use, but by simplifying the machinery under it, also much cheaper. You will judge of the utility of my invention by an experiment I have already made. In traveling from Paris to Naples, we were three of us, with much baggage in my coach: my springs were English, and of the best quality: though we traveled post with six, and sometimes eight horses, over paved and broken roads, sometimes hard frozen, they never absolutely broke, but were constantly giving way: sometimes three or four plates would crack, sometimes the iron that supported them would break, and at other times they would tear and wreck the wood to which they were fastened: scarce a day passed that we were not compelled to have some repairs made, though we strengthened them with cords and thin slips of wood, as much as possible. On my return from Naples, they underwent a complete repair at Rome; the defective plates were taken out, and new ones put in; they were covered with wood, and the whole carefully corded; a precaution, without which no iron springs will stand traveling post a thousand or fifteen hundred miles; particularly as the postillions, instead of having any mercy upon them, do all in their power to break them. When they enter or leave a town or village, the pavement of which is extremely broken, they snap their whips in such a way as to bring all the inhabitants to their doors and windows, and put their horses upon full gallop, to show their address in driving. Before I got to Bologne, I found new repairs necessary, and I began to fear that no repairs would enable me to complete my journey through Germany with the same carriage. This determined me to try the following experiment:—

At Bologne they make foot-balls of asses' skin dressed in oil, and containing some oil to keep them supple. I purchased four of these, and after covering them with calf-skin, placed them between the two folds of the thorough braces, behind and before, where the screw springs are sometimes placed. These exceeded my expectation. Though I traveled in the months of February and March, when the roads were at their worst, through a considerable part of Italy, through the Tyrol and Germany, and through the paved roads of France by the way of Strasburgh to Paris, a journey of many hundred miles, not a spring gave way, nor did any part of the carriage break; though I found before I arrived at Munich, that the air had escaped from one of the balloons that was placed under the front springs. The motion of the coach was also much easier than it had been before the application of the foot-balls.

Perhaps springs of this kind might be adjusted to saddles, so as to render the motion of a hard trotting horse as easy as that of a Narragansett. Air-cushions would be admirably adapted to the seat of the common Dutch wagon. These might, perhaps, be made out of the stomach of an ox or horse, well tanned and dressed in oil, and blown up to $11^{1}/_{2}$ atmosphere, or 22 lbs. pressure upon a square inch. Nor could a lighter or warmer coverlet for beds be contrived, than silk, rendered by elastic gum impenetrable to air, and stuffed with that material. I do not think it impossible even to make beds of it. And I sincerely wish it was effected, if it was only to relieve our poor geese from the horrible torture our luxury makes them undergo.

Professor W. A. Norton, of Delaware College, Newark, Del., presented and read a communication, entitled, "An Inquiry into the Constitution and Mode of Formation of the Tails of Comets."

The two distinct topics embraced in the title of this paper, were discussed in the order in which they are there named. In treating of the first topic, it was taken for granted, inasmuch as it is admitted by all astronomers, that the tail of a comet consists of the same kind of matter as the head; and accordingly is not a mere spectre of light, as some theorists have advocated. It was then argued, that the tail was formed from the matter of the nebulous envelope of the head: and in the same connection it was incidentally established, that the particles of the tail could not revolve disconnectedly in separate parabolas, by showing that the tail would not on this supposition continue in the position of opposition to the sun; also that, supposing the particles to be connected in any way whatsoever, the situation of the tail could not be accounted for without taking into view the action of some force other than the sun's attraction.

Under the second head it was shown—1. That the sun is concerned, either directly or indirectly, in the process of forming the tail of a comet. 2. That the particles of matter which make up the tail, must have been driven off from the head by some force exerted in a direction from the sun. 3. That this force cannot emanate from the nucleus. 4. That it must therefore be a force taking effect upon the matter of the comet from without, and from the sun outwards. 5. That it extends far beyond the earth's orbit, developing tails there as well as near the sun. It was inferred, from analogy, that this force decreased in receding from the sun. No attempt was made to investigate its nature; but since it acted precisely like a repulsive force, it was called, for the sake of simplicity of conception, the *sun's repulsive force.* Such being the force conceived to be the efficient cause of the development of the tail, there are two modes in which it may be supposed to act, viz.—1. By expelling a certain portion of nebulous matter into the form of a tail, which, being once fashioned, remains the same, revolving along with the nucleus until more matter chances to be driven off to add to its brightness and extent; which has been the received notion hitherto; or—2. By *continually* urging a portion of the cometic matter away from the head an indefinite distance into free space; in which case the tail, as seen by us at any instant, would be but the collection of all the particles that had been emitted from the head during a certain previous interval, viewed in the act of darting off into free space. This theory, so far as known, has not been propounded before by any astronomical writer. Agreeably to this notion, the tail increases in length towards the perihelion, by reason of a more copious evolution of shining nebulous matter, in consequence of the increased heat of the sun. It diminishes in brightness in receding from the head, and at last becomes too faint to be discerned, from the following causes:—1. Amore rapid flow of the matter by reason of a longer continued action of the sun's repulsive force. 2. An increase in the breadth of the tail. This may be supposed to arise from the divergence of the lines of direction of the forces acting upon the outer parts of the envelope, and, in some cases also, from a rotation of the tail about its axis, generating a centrifugal force. The tails of some comets are known, from observation, to have had a motion of rotation about their axis, as the comets of 1769 and 1825, and Halley's comet at its last appearance. 3. An augmentation in the distance of the matter from the sun, the supposed source of its light.

The following objections were then urged against the received theory:—1. No good reason can be assigned why the sun's repulsive force, so called, should not drive off the nebulous matter of the tail still farther; since it extends indefinitely into space, and has sufficient energy to expel comparatively dense matter immediately from the nucleus of comets, that are more remote from the sun than the extremities of the tails of the comets, which have the smallest perihelion distances. 2. This being true, the repulsive force, which alone can be supposed to keep the tail in rotation so as to be always opposite to the sun, ought to dissipate it, rather than set it in rotation as one connected mass. 3. Agreeably to this theory, the tails of all comets that come near the sun, ought to be completely dissipated by the centrifugal force, in passing round this luminary; whereas no such fact has been observed. This was illustrated by the case of the comet of this year. It was shown, that on the supposition that the tail revolved with the nucleus, and at the same time rotated at such a rate as to keep opposite to the sun, the extremity of the tail must have had the amazing velocity of at least twenty-one thousand miles per second; that the centrifugal force must have been a hundred times greater than at the nucleus; and the gravitation toward the sun ten thousand times less.

It was next attempted to account for the phenomena of the situation of the tail, viz.—1. its general situation: 2. its deviation from the position of exact opposition to the sun, which is first perceived at a certain distance from the perihelion; after which it increases progressively until a short time after the perihelion passage, when it begins gradually to diminish, and finally becomes insensible: 3. the curvature of the tail. The first mentioned phenomenon is a simple consequence of the theory: the second also. It was shown, that the particles being driven away to any distance from the nucleus, still retaining their velocity in the direction parallel to the orbit, would at the end of any interval be in such situations, that the lines joining them with the nucleus would be parallel, respectively, to the radii vectors, along which produced they were emitted from the nucleus; from which it followed, that the tail would deviate everywhere from the radius vector prolonged beyond the nucleus, and at the same time be curved and concave toward the regions of space which the comet has left. But this deviation would only be perceptible in the vicinity of the perihelion where the motion in anomaly was the most rapid. The smallness of the curvature throughout the greater part of the length of the tail was explained by supposing that the velocity of emission soon came to be very great in comparison with that of the motion in anomaly. It was conceived, that the greater curvature at the extremity might arise from a diminished velocity of flow, by reason of the resistance of an ether in space, in conjunction with the falling back of the particles, in consequence of the resistance of the ether to the motion parallel to the orbit.

Finally, the objection to the theory that had been exposed, of a continual waste of the material of the tails, and a consequent gradual diminution in the length and brightness of these appendages, until at last they are wanting altogether, was considered. It was argued, that there were now many comets entirely devoid of tails, and that these may have already experienced the fate that is supposed to await all the others. Also, that multiple tails spring up suddenly, and then vanish in a few days. That the principal tail is more durable than these, because the stock of materials from which it is derived is larger, and is frequently supplied from vapor rising from the nucleus. The cases of Halley's comet and others were referred to in proof of this last assertion. It was added, that the waste here supposed was believed by many astronomers, from the results of observation, to be in actual progress.

Professor Draper presented a communication, entitled, "On the Decomposition of Carbonic Acid and the Alkaline Carbonates by the Light of the Sun," by John W. Draper, M.D., Professor in the University of New York City.

For many years it has been known that the green parts of plants, under the influence of the sunlight, possess the power of decomposing carbonic acid and setting free its oxygen. It is remarkable that this, which is a fundamental fact in vegetable physiology, should not have been investigated in an accurate manner. It is not known that any one has yet attempted an analysis of the phenomenon by the aid of the prism, the only way in which it can be truly discussed.

It is the object of Prof. D.'s paper to prove, 1st. That the light of the sun is the true cause of the decomposition, the rays of heat and the so called "chemical rays" not participating therein, as Graham, Johnston, and other writers on vegetable chemistry suppose. 2d. That it is the yellow light, or most luminous ray, that is mainly concerned. 3d. That leaves evolve not pure oxygen gas, but a mixture of oxygen and nitrogen in regulated proportions. 4th. That there is an extensive class of salts which is decomposed under the same circumstances, and therefore the phenomenon is rather to be attributed to a digestive than to a respiratory process. 5th. That this digestion is brought about in the same way as the digestion of animals, by the decay of a nitrogenized body.

To show that the light of the sun is the cause of the decomposition, having obtained a motionless spectrum by the aid of a heliostat, he placed in the different-colored spaces tubes filled with water impregnated with carbonic acid gas, and containing some leaves of grass. The decomposition presently commenced, and in the course of two hours a sufficient quantity of gas was collected. On examination, it was found that the tubes in the yellow, the orange, and the green light, contained most gas; that in the red, a much smaller quantity; and those in the blue, the indigo, and the violet, none at all.

But the maximum of heat occurs in or beyond the red ray; the maximum of chemical action among the more refrangible colors, blue, indigo, and violet; and in these spaces the decomposition of the acid fail to go on.

From this he infers that it is the light of the sun, and the yellow light mainly, that is the cause of the phenomenon.

On causing leaves to decompose carbonic acid in water by the rays of the sun, and collecting the gas as it is evolved, it appears on no occasion to be pure oxygen, but a mixture of oxygen and nitrogen in variable proportions; from fifty to ninety per cent of oxygen being found at different times, as is shown by explosion with hydrogen gas. But although there is this great variability in the proportion of the two gases evolved, a very simple law, which directs the progress of the decomposition, may be traced. On causing leaves to decompose a known volume of carbonic acid, the same volume of the mixed oxygen and nitrogen makes its appearance. From this it is to be inferred, that plants during this action do not only effect the fixation of carbon, as is commonly supposed, but with it they absorb a certain amount of oxygen also. When a leaf, exposed in carbonic acid gas to the sunshine, has completed its function, it has appropriated or assimilated all the carbon, and with it a certain portion of the oxygen; the residue of the oxygen has been evolved, and with it a volume of nitrogen precisely equal in amount to the volume of oxygen appropriated by the plant.

This disappearance of oxygen and appearance of nitrogen are thus connected with each other: they are equivalent phenomena.

The emission of nitrogen is not a mere accidental result, but is profoundly connected with the whole physiological phenomena.

The elementary conditions under which carbonic acid gas is decomposed having been thus stated, Prof. D. passed next to the description of similar decompositions occurring in the case of saline bodies. It has always been a subject of surprise to chemists, that the powerful affinity which carbon and oxygen are thus held together, should be so easily overcome at common temperatures. Even potassium cannot decompose carbonic acid in the cold. It might therefore be reasonably expected that the energetic forces which bring about this change ought also to effect other remarkable decompositions. In fact, the decomposition of carbonic acid is only one of a very numerous series.

Having boiled some distilled water to expel all gaseous matter, dissolve in it a small quantity of bicarbonate of soda. Introduce into a test tube some leaves of grass, fill the tube with the saline solution which has been once more boiled to expel any air it may have obtained from the dissolving salt, and invert the tube in some of the solution in a wine glass, after having carefully removed all adhering bubbles of air from the leaves by a piece of wire, or in any other convenient manner. This arrangement, kept in the dark, undergoes no change; but if brought into the sunshine, bubbles of gas are rapidly evolved, and in the course of a few hours the tube becomes half full. On detonation with hydrogen, this gas proves to be rich in oxygen.

Prof. D. made some attempts to discover how much oxygen could in this way be evolved from known quantities of bicarbonate of soda; supposing it probable that the second atom of carbonic acid being removed and decomposed, the process would cease: however, the results of his experiments indicated that the supposition he had formed was not correct. The process is not limited to the removal and decomposition of the second atom, but goes forward, the first itself being in like manner decomposed. From this it would seem that carbonate of soda itself should be decomposed; and experiment verified the conclusion: for on using that salt instead of the bicarbonate, the evolution of oxygen went on precisely in the same way.

As in these experiments a solid salt is decomposed, it is obvious that the function by which the leaves accomplish this is very different from that of respiration. It is not respiration, but a true digestion.

In the same way Prof. D. found that all kinds of soluble carbonates and several other organic salts, such as bitartrate of potash, citrate of soda, succinate of ammonia, &c., would emit oxygen.

It thus seems that the decay of some nitrogenized body in the leaf is essential to the digestive action of plants.

At this stage of the inquiry, a remarkable analogy appears between the function of digestion in animals and the same function in plants. Liebig has shown, how from the transformation of the tissues of the stomach itself, food becomes acted upon, and is turned into chyle, an obscure species of fermentation brought about by the decay of nitrogenized bodies. So in like manner in plants, the dissolution of a nitrogenized body brings about the assimilation of carbon. The facts seem to indicate that the primary action of the light is not upon the carbonic acid, but upon the nitrogenized ferment; and that the decomposition of the gas occurs as a secondary result. From this we may infer that chlorophyll, the green coloring matter of leaves, is the body which in vegetables answers to the chyle of animals; that it is derived from the decomposed carbonic acid through the eremacausis of albumen brought into the leaf, or of some compound of the elements of ammonia that passes up by the route of the ascending sap; and that the oxygen which disappears, disappears to bring about the eremacausis of that ferment. Under this point of view, the digestion of plants may be regarded as taking place in the following way: There is introduced into the leaf some azotized body formed by the aid of ammonia that has passed through the spongioles: on this the sunlight acts, bringing about its decomposition by causing its union with oxygen: and now, if carbonic acid be present, the decomposition is propagated to its atoms; a part of the oxygen set free is expended in continuing the eremacausis of the ferment; the rest is evolved with an equivalent volume of nitrogen. The carbon thus set free unites at once with the elements of water, and chlorophyll results. But this chlorophyll undergoes continuous change under the action of the sun, and is as continually replaced; from it is formed gum, and finally lignin, and all the woody fiber of plants must have originally existed as chlorophyll, or passed through the green stage.

SPECIAL MEETING.

Fifth Session, 29th May, 10 *o'clock, A. M.*

DR. BACHE, Vice-President, in the Chair.

Letters were read from Professor Bartlett, of West Point,—
Professor Caswell, of Brown University,—and Professor
Locke, of Cincinnati.

Dr. Samuel George Morton, of Pennsylvania College, read a
"Summary of his Series of Observations on Egyptian Ethnography."

Dr. Morton stated, that having from time to time communicated
the result of some inquiries into Egyptian ethnography, he took
the present occasion to present a summary of his observations, in
connection with a part of the data from whence they have been
derived. Dr. M. then submitted to the inspection of the Society
twenty embalmed heads from the Egyptian catacombs; being part
of a series of one hundred, sent him by George R. Gliddon, Esq.,
during the period in which that gentleman held the United States'
Consulate at Cairo. They were obtained from seven different sepul-
chral localities, upwards of six hundred miles apart, beginning at
the Memphite Necropolis in the north, and terminating at the
ruined temple of Parembole, in Nubia.

Dr. M. then observed, that a careful analysis of these remains
had enabled him to refer them to two of the great races of men,
viz. the Caucasian and the Negro, although there is a remarkable
disparity in the number of each. Again, the Caucasian skulls vary
so much among themselves, as to present several different types of

this race, which have been heretofore mentioned in the Society's Proceedings for November and December 1842.

1st. The *Pelasgic form* of the head, or that which most nearly resembles the conformation of the people of Western Asia and Southern Europe. The Pelasgic head is familiar to us in the beautiful models of Greek sculpture, which are remarkable for the volume of the head compared with that of the face; the symmetry and delicacy of the whole osteological structure, and the largeness of the facial angle.

2d. The *Semitic form,* as seen in the Hebrew communities, has been characterized from remote antiquity by a receding forehead, long, arched, and very prominent nose, a marked distance between the eyes, and a strong development of the whole facial structure.

The number of heads conforming to the Pelasgic and Semitic types,—for which Dr. M. on a former occasion suggested the provisional name of *Arcto-Egyptian* group,—is fifty-one, or a fraction more than one-half of the whole series. It is moreover remarkable, that the physiognomy of the ruling caste of the Egyptians, as seen in the portraits yet preserved in the paintings and sculpture of their temples, corresponds in a remarkable manner with the character of the head and face of the Pelasgic series of mummies. Thus the portrait of Amunoph 1st, the oldest which has hitherto been identified, presents the finest cast of European features; and yet this monarch reigned in the valley of the Nile, and held his court in Memphis, more than eighteen centuries before the Christian era: and if from this remote period, we trace the physiognomy of the kings and queens of the subsequent reigns, we perceive among them many equally beautiful models of the human countenance, some of which are not inferior to the beau-ideal of classic art.

3d. The *Austral-Egyptian* group embraces that modification of the Caucasian type which is seen in the Hindoos and Southern Arabs, blended with some peculiar traits which may have pertained to the aboriginal inhabitants of Meroe. The head is smaller than in the Pelasgic type, and the forehead narrower; while the face being more prominent, the facial angle is rather less. The nose is straight or aquiline, the face angular, and the features often sharp. This modification is also frequent on the Egyptian monuments, occasional, for example, in the ruling caste, but very frequent among the subordinate classes.

Dr. Morton next proceeded to speak of the *Negro* race, of which but a single unmixed and unequivocal example occurs in the entire series; and this one is not from any of the catacombs, properly speaking, but from a tumulus proximate to Philm, on the frontiers of Nubia; a spot which is supposed to have been a receptacle for the people of all nations who chanced to die while in attendance at that celebrated shrine. But although the unmixed Negro occurs but once in the series, there are eight other heads which possess decidedly blended characters in which those of the Negro predominate. For these the author has proposed the name of *Negroloid* heads; for while the osteological development and the very *expression* of the bones are that of the African, the hair is long and rather harsh; thus indicating that combination of features which is familiar in the mulatto grades of the present day, and which is also characteristic of a part, at least, of the modern Coptic population of Egypt.

After illustrating these varied conformations by means of the embalmed heads already alluded to, Doctor Morton submitted a revised ethnographic table of the whole series of Egyptian crania, according to the *preponderance* of physical or organic characters, as follows:—

1. Heads of the Pelasgic type, ... 44
2. Heads of the Semitic type, ... 7
3. Heads of the Austral-Egyptian type, ... 33
5. Negroloid or Coptic heads, ... 8
6. Negro, ... 1
4. Austral-Egyptian heads, with traces of Negro
    lineage, and some other but indeterminate traits, ... 5
7. Idiotic heads, race uncertain, ... 2

———

100

———

He then submitted the following additional observations derived from the materials in his possession :—

1. The internal capacity of the cranium, (which of course indicates the size of the brain,) measures ninety-seven cubic

inches in the largest Caucasian head, and sixty-eight in the smallest; and the average derived from the three divisions of this race will not, perhaps, exceed eighty-five cubic inches. The few Negroloid heads give an average of eighty cubic inches, and the single unmixed Negro but seventy-one. Thus the brain, in each series, presents a smaller average size than is characteristic of any of the corresponding races of the present day.

2. The Facial Angle gives the following measurements:—

Average of the Pelasgic series, ...80°
Average of the Austral-Egyptian series,.................78°
Average of the Negroloid series, ...75°

3. Position of the Ear. The singular elevation of the ear, so familiar in Egyptian statues, bas-reliefs, and paintings, is not confirmed, as a natural organization, in this series of heads. The bony meatus presents no deviation from the usual relative arrangement of parts; but the *cartilaginous* structure being desiccated, and consequently contracted, may not afford satisfactory evidence.

4. The observation of the late venerable Professor Blumenbach in respect to the *incisor teeth* of the Egyptians,—viz. that, in comparison with those of other nations, they were large, thick, and cylindrical, or obtusely conical, instead of having the characteristic chisel-like form,—is not confirmed by the present investigation. For among upwards of two hundred teeth of this class, derived from forty-five different crania, there does not appear to be any deviation from the ordinary form or structure; and the conical appearance of the crowns of the teeth has obviously resulted from the habitual mastication of hard substances, and is common in the Hindu and other nations, in whom it is derived from like causes.

5. The Hair is preserved on thirty-six heads, in some instances in profusion, in others scantily, but always in sufficient quantity to permit an examination of its texture. Thirty-one pertain to the Caucasian race, and in these the hair is as fine as in the fairest European nations of the present day. Of the eight Negroloid heads, those which retain the hair have it

comparatively coarse, but not woolly, and in one instance it is nearly a foot in length. In the solitary Negro head, the hair is too characteristic to be mistaken.

6. These materials, in conjunction with monumental evidence, go to establish the antiquity of those physical or organic characters which constitute the differences among the several races of men; and show us, by direct and unequivocal contrast, that these differences are as old as the oldest records of our species.

7. The origin of this extraordinary people has hitherto been one of the most difficult problems in ethnography; and this difficulty has chiefly arisen from the fruitless effort to trace to a single source that nation, which derived its origin, as we have seen, from several distinct branches of the human family.

Mr. Job R. Tyson read a paper on "The Social and Intellectual State of the Colony of Pennsylvania, prior to the Year 1743."

In this essay, Mr. Tyson considers the social condition of the founders and early inhabitants of Pennsylvania, and traces their progress in scientific and literary pursuits up to the organization of the American Philosophical, Society, in 1743. It is evident, he says, that such an association as this must have been the result of preexisting causes, and that the success which has followed it must be less owing to the happy or fortuitous circumstances which attended its birth, than to the steady operation of other influences, which were coeval with the establishment of the English province.

The subjects of Gustavus Adolphus, who, before the middle of the seventeenth century, alighted on the shores of the Delaware, seem to have been a frugal, honest, and worthy race; but I cannot find, said Mr. T., that either they or their Dutch invaders paid much attention to the interests of learning. The colony of New Sweden was small in number, the inhabitants extremely illiterate, and its social state one of unattractive rudeness, of unalloyed but rustic simplicity.

The English settlement by Penn was more numerous, and projected with loftier aims. It occurred at a propitious period, and under circumstances favorable to the development of a healthy national character. The civil wars of England, and the great rebellion in which they terminated, were past. The fury of religious persecution was stayed, and the heat of religious controversy, though still excited and feverish, was not as before to be quenched by blood. A new order of men had arisen out of the burning cauldron of puritanism, which, though partaking of the puritan leaven, was tempered by cooler heads and milder tenets. The Quaker sect, at the head of which stood Robert Barclay, William Penn, George Whitehead, and others, proclaimed to all—even to the hunted Jew and proscribed Mahometan—the novel doctrine of universal toleration, and united with this sentiment a great variety of opinions, deemed subversive of existing dogmas, and threatening the privileged orders of the Church and State of England.

In order to reduce some of these principles to practice, Penn accepted in 1681 a charter of Pennsylvania from Charles II. Thither he repaired in the succeeding year with such companions and followers, as animated by the hope of improving their decayed fortunes, or induced by the adventurous spirit of change, or anxious to enjoy their religious tenets freed from the oppressive or uplifted hand of secular authority, were willing to encounter the austerities of a residence in the new and remote regions of the west. Here the liberal principles of the founder were to stand the trial of experiment. The problem was to be solved, whether government, exposed to the billows and inundations of the democratic element, and subjected to the dangers of unfettered religious opinion, could subsist without the nutriment of a hierarchy, without the distinction of caste, and without the aid of privilege.

As all religious professors were equally entitled to protection by the great law of 1682, multitudes flocked to the new settlement. But notwithstanding the freedom which was allowed to discussion and conduct, and the constant influx of strangers from England and the neighboring colonies, it does not appear that religious controversy engaged much of the Colonial mind, or that with the exception of the Keithian schism, diversities of sentiment estranged the affections or excited the passions of the people. The minds of the settlers, thus left free to think and act without the apprehension of restraint, or the dread of a superior, directed their powers fearlessly to the question of government, to the melioration of their physical state, and to the improvement of their moral and intellectual condition.

> —nec verba minacia fixo
> Ære legebantur; nex supplex turba timebat
> Judicia ora sui.—Ov. MET.

The early emigrants included in their number men of good educations and high endowments. Penn himself was a scholar and a writer; his mind was of a sagacious and original order, and enriched with various and profound knowledge. Thomas and David Loyd, Makin, Pastorius, Kelpius, Hamilton, Logan, Norris, Brooke, Keith, and many others who could be named, were men of considerable classical attainments. It is enough to say, that the mathematics and ancient languages were taught in the Friends' Public School; that the genius, scenery, and peculiarities of the province were soon celebrated in Latin verse; and that the Roman and French tongues were, on one occasion at least, resorted to as the mediums of intercourse between the English and German emigrants.

A printing press was in operation in Philadelphia, so early as the year 1686. This was only four years after the settlement by Penn, while the forests were standing in primeval wildness around the colonists, and before huts were substituted for the caves which first sheltered them from the inclemencies of winter. In all the other colonies, this engine of mind was postponed till the asperities of a new country were subdued by longer cultivation, or until physical ease gave more leisure to seek for mental conveniences. In Pennsylvania, the cause of education and the diffusion of knowledge by means of printing, were contemporary with the landing. The following year (1687) is signalized by the printing of an *almanack*. This performance was from the printing house of Bradford, and is remarkable as one of the first emanations of the Colonial press. In conformity with a provision in the frame of the Government, a school was opened in the next year after the landing (1683), and in six years afterwards was established a Friends' Public School, where the poor were taught gratis, and sound literary and scientific learning was open to all. The preamble to the charter, which was granted to this seminary in 1701, shows the high aims of the colonists with respect to mental culture. It recites that the prosperity and welfare of a people depend mainly upon the good education of youth, and that the qualifications for public and private usefulness are chiefly derived from learning to read and write, and from "the learning of languages and useful arts and sciences, suitable to their sex, age and degree," &c.

James Logan accompanied the proprietary on his second visit to the colony, in 1699. His valuable treatises in Latin, and his English translation of Cicero's little work, *De Senectute*, are well known. These have given to posterity additional evidence, if any were wanting, of his devotion to literature and science. With great liberality, he bequeathed the books known as the Loganian collection of the Philadelphia Library to the city, "for the advancement and facilitating," as he observes, "of classical learning." He was fifty years in forming this library, which numbered nearly four thousand volumes at his death. It included one hundred folio volumes, in Greek, mostly with versions. The Roman classics were among them, "without," he says, "an exception." All the Greek mathematicians, Archimedes, Euclid, Ptolemy, &c., had a place, besides a great number of modern mathematicians. In addition to standard works of enduring value, many rare and curious volumes are to be found in this collection, which, at the present time at least, to use his own expression, "neither prayers nor price could purchase."

In the year 1719, the first newspaper was published in the colony of Pennsylvania, under the title, "The American Mercury." The Boston News Letter, undertaken and published in the year 1704, at Boston, by John Campbell, a Scotchman, claims the undeniable distinction of being the first newspaper which appeared in either of the North American colonies. Though Pennsylvania, which is half a century younger than Massachusetts, must yield this honor to her elder sister, yet the priority is a period of only fifteen years, and at Philadelphia was published the first *daily* newspaper which appeared on the continent.

Four years after the commencement of "The American Mercury," Franklin appeared, a poor and friendless boy of seventeen, in possession of a trade about half taught, in the streets of Philadelphia. Before I refer to the history of this remarkable man, or the effects which his presence and exertions produced upon our institutions, it may be proper to show how circumstances contributed to his success.

We have seen that the leading minds of the first settlers were scholars; that the means of common and scholastic education were amply provided; and that zeal and enterprise in the cause of learning were exhibited in the early establishment of a printing press, and in a variety of literary performances. It remains to be shown, that the principles of the Colonial policy had concurred with these causes, in diffusing a self-respect and spirit of generous rivalry among those classes of society, to which in other countries they were strangers.

Among the beneficial influences which the Society of Friends exerted upon the infant colony from its establishment, were the recognition of usefulness in occupations, simplicity in living, and equality in classes. As these principles were engrafted in the maxims of their religious profession, they taught that each was to be deduced as a corollary from the humility of the Christian character. The callings of men however humble or laborious, were not permitted to detract from their social standing; and frugality in living and simplicity in furniture and dress were enjoined on all their members, without reference to their pecuniary means of indulgence, or their taste for luxury or expense. Those arts which merely embellish life, and add to our enjoyments in the gratification of the senses, were decried. Nothing was deemed meritorious, or voted to be respectable, but that which could be made subservient to the great purposes of utility or practical convenience.

They taught that as trades and manual labor were useful, assiduity in their prosecution was honorable. William Penn recommended trades to his children. Other leading Friends, whose ancestors, claiming for the most part a cavalier descent, had always belonged to the best classes of English society, adopted his sentiments, and set the example of bringing up their children to some useful or handicraft employment.

The necessities of a new country gave force to these suggestions, and induced their extensive adoption by the colonists. The effect of such views upon a society, in which existed no titular ranks, except those which must result from the inevitable subordinations of social and political life, was pervading. The principle had its origin in religious faith, and that only, without looking to political consequences. While this principle left the claim of conventional honor untouched, it raised to respectability a class of men, whose ignorance and occupations had before consigned them to the evils of neglect and a chiliing sense of inferiority. Birth and employment came to be disregarded in the estimate of personal character. However humble and depressed both might have been esteemed elsewhere, they presented no obstacle in Pennsylvania to advancement and consideration. Perhaps no event in history has tended so much to the real elevation of the working classes as the religious maxims and social scheme of Penn and his companions in the Province of Pennsylvania.

All this had the salutary effect of bringing the different classes of society into closer union. The social manners of mechanics, condemned in England to isolation, were improved; and their prevailing sympathies and impulses softened and enlarged. They were soon taught to feel the advantages of scientific knowledge to the manual arts, and to see the connection subsisting between them. The mechanic of Pennsylvania thus became a different sort of person from the mechanic of other countries. Many of her practical farmers and unambitious tradesmen were the offspring of refined and educated parents, who, in training the hands of their children to labor, did not forget the cultivation of their minds, nor the improvement of their religious and moral faculties.

So universal was the adoption of trades in the colony, either with or without some other employment, that few exceptions occur until after the middle of the last century. The placid surface of the social stream became disturbed in the tumults of the revolution, and in the upheavings caused by war, the filth and deposits of the current, whose natural resting place was the bottom, sometimes mounted to the top. It was thus that social as well as political life underwent a change.

Tantae molis erat Romanam condere gentem.

ÆN.

In the excitements of a momentous contest, in the more enlarged views which its successful issue presented, in the rivalry and competitions for political office, and in the diffusion of more luxurious tastes and habits, the primitive ideas of devotion to practical husbandry and the mechanic arts gave place to the engagements of commerce, and thence to the more ambitious and ornamental pursuits of life. But the principle in its native integrity was preserved, and is still exemplified by many members of the religious sect in which it originated.

In connection with the ideas of frugality, simplicity, and utility inculcated by the first colonists, it must not be forgotten that they were equally diligent in cultivating the benevolent principles of man, which they sought to awaken by private opinion and nurse by the stimulus of keeping them in constant exercise. The value of physical means, appliances, and instruments in the government of the world, was depreciated; the animal instincts and propensities were to be subdued, if not extinguished. In pursuance of this scheme, they denounced war and fighting: they condemned the severity of the lash and other modes of physical torture in the punishment of offenders; and declaimed against capital inflictions. Instead of these, they set about mitigating the rigor of the penal code: jails were reformed and meliorated, and charities founded for the poor and unfortunate.

The system of African slavery found no support, and as practiced, no sympathy nor encouragement from William Penn; and his brethren of the province, after long continued and ineffectual remonstrance, finally determined, in the early part of the eighteenth century, to exclude from religious fellowship such of their members as were concerned in the traffic. Pennsylvania owes to her Quaker colonists,—especially to her founder,—to Southeby, Sandiford and Lay,—to Elisha Tyson, Anthony Benezet, and John Woolman,—the worthy distinction of setting an example to the other states of the Union, of so modifying her system of domestic servitude as to bring about, in a few years, its gradual but final extinction. This memorable event took place in the year 1780.

Whatever may now be thought of some of the theories advanced by the Quaker puritans of that day, it must be admitted that ideas growing out of reflections upon our moral being, and based upon the improvable capacities of our moral nature, could only spring from minds enlarged by study and refined by general cultivation. It is to these causes we owe the numerous charitable foundations for which Pennsylvania is so honorably distinguished, and the celebrity she has long enjoyed for her mild punishment of offenders, and the latest improvement of her penitentiary system.

It is not surprising therefore that Franklin, on his arrival in Pennsylvania, should find apprentices whose aspirations were equally generous with his own. When he founded with characteristic sagacity that remarkable union, the Junto of 1727, those who sympathized in his project were mostly mechanics, and brought up in the same sphere of life with himself. The members of the association were to be confined to twelve in number, but according to Franklin's account, the original number of those who were actually enrolled, was eleven. Of these, Thomas Godfrey was a glazier, William Parsons, a shoemaker, William Maugridge, a joiner, and Hugh Meredith and Stephen Potts "were bred to country work." but at that time, the former was engaged "to work at the press," and Potts was at bookbinding. Of the other five, Joseph Breintnall was "a copier of deeds for scriveners"; Nicholas Scull was a surveyor; George Webb is described as "an Oxford scholar," but his time, for four years, had been purchased by Keimer, the printer; William Coleman was then a clerk, and Robert Grace was a young gentleman of some fortune. These, with Franklin himself, the author of the society, who had been struggling with penury as a journeyman, but who now was a *master* printer, comprised the company. The promiscuous association of different classes, as displayed in the occupations of the members,—classes, which, in Europe, had seldom come into contact with each other,—cannot escape notice. No doubt the social fusion which it evidences, was promoted by the commanding intellect of the man who planned the enterprise; but more certainly may be ascribed to the amalgamating properties of other and antecedent elements. The notion of transmitted and hereditary virtue, however we may condemn it as absurd and unphilosophical, cannot be overcome by suggestion, or obliterated in a few years. In Pennsylvania, the original structure of the social state had been placed upon new foundations, and leaned for its support upon reason and principle, not upon the fallacies and delusions of prejudice or the maxims and examples of antiquity.

The members of the original Junto were ingenious men, whom the love of knowledge had assembled, and whom the most generous aspirations cemented together. I will not repeat what is so generally known respecting their characters and attainments; as the delightful autobiography of Franklin himself, who has characterized each, and the volumes of the Historical Society of Pennsylvania, furnish very copious information.

From the ingredients of the Junto, as well as from the contents of the Logan library, it is evident that light literature and graceful verses did not absorb the mind of the province. The satire of Young against a pursuit of the muses had appeared, and though it was caustic enough for so poetical a temperament, it could not eradicate a taste already formed. Mr. J. F. Fisher has shown that many of the colonists cultivated the muses with very tolerable success. But the tendency of the Colonial mind was to useful acquisitions in science. This arose from the convictions of our ancestors, already
referred to, that the elegant and ornamental arts were worthy of little encouragement and care. Education was too generally disseminated to permit the extinction of a classical taste; but though versifiers occasionally appeared, and a love of light literature was widely diffused, yet the energies of the youthful province were reserved for pursuits more congenial with practical exigencies and the predominant feelings of the country.

We have seen that the Junto was formed in the year 1727. In the following year, Makin wrote his Latin poem, entitled, "*Encomium Pennsylvaniæ*" to which succeeded, in the year 1729, his "*Descriptio Pennsylvaniæ.*" These verses are not without merit as metrical compositions, and show at least that the author had studied the classical productions of Rome, and understood the structure and prosody of the Latin tongue. The Library Company of Philadelphia was started two years after, in 1731, and had its origin, under the auspices of Franklin, in the desire of the Junto to have a permanent collection of books for the benefit of its members. This Library, it may be observed in passing, though now unequal to the literary wants of Philadelphia, has risen to an importance far exceeding in number and value any other bibliothecal repository in the United States. It certainly argued a diffusive zeal for knowledge, that in an infant and sparsely populated colony, fifty original subscribers, and they "mostly among young tradesmen," could be obtained for such an enterprise, with the expectation that an annual contribution would be required for the space of half a century. In 1741 was attempted a Magazine, which is the first effort in any of the colonies to establish a literary journal. In the following year another newspaper was established in Philadelphia. About this time it was that James Logan published at Leyden several works in Latin on different branches of science, and in the province his English translation of Cicero on Old Age; that Thomas Godfrey, the author of the quadrant, was prosecuting his ingenious and scientific labors; and that John Bartram, whom Linnaeus justly pronounced the greatest natural botanist in the world, was earning honor from his sovereign, and fame from the learned societies of Europe. These and kindred occurrences prepared the way for further events. In the year 1743 an Academy was suggested, which grew into a great literary and medical university, whose well-earned and unrivalled eminence has long been a source of cherished and honorable pride to Pennsylvania; and the same year witnessed the formation of the Philosophical Society, whose centennary we have just celebrated.

Many original works were published before the era of 1743, of which a considerable number is still preserved in the city Library. There are now on the shelves of that institution above four hundred original books and pamphlets, which were issued by the Philadelphia press before the revolution. A multitude of domestic productions are no doubt lost, and if we add the reprints of foreign books, in which, at all periods, our press was prolific, the number of works printed in the colony may be estimated much beyond what is generally imagined.

The aim of this paper, said Mr. Tyson, is accomplished, in showing that in the year 1743 the formation of such a society was not forced or premature, but that amid the general culture and scientific predilections of the colony, it was as natural, as it was certainly important, to combine and, concentrate intellectual exertion. Like the other institutions, which the mental wants of the country demanded, it became itself the nursing mother of our infant science and the great distributor of its scientific wealth.

Mr. Tyson closed his essay by sketches of the character and habits of a few of the more distinguished members of the Society of 1743.

Dr. Meigs read a communication, entitled, "Measurements of the Fœtal Cranium, by Charles D. Meigs, M.D., Professor of Obstetrics, &c. in the Jefferson Medical College."

He stated it to be universally admitted by medical writers, teachers, and practitioners, that it is important to know the dimensions of the head of the child at birth. Without correct information upon this point, the operations of the surgeon would be still more difficult, dangerous, and uncertain, than they necessarily are even with the most accurate knowledge of the form, size, and resistance of the fœtal cranium, and the relative magnitude of the pelvis.

Having been himself employed for some years in this department of surgical pursuits, he had long supposed that the estimates of writers and teachers upon the point in question were less accurate than they ought to be; since he had observed the size of the cranium to be greater than is generally allowed.

Dr. Meigs was not prepared to say, that the calculations made in other places might not be just for those places; but he was convinced that they are too low for Philadelphia, and probably for the United States. He knew of no measurements taken in this country, and therefore hoped that his might be not unacceptable to the Society.

Having provided himself with what is called a pillar compass,—so called by the trade,—and which is a most convenient and portable calliper, he applied it in succession to the heads of one hundred and fifty children born at term under his care. After using the compass as a calliper, the points were pricked into a slip of paper, and then, being carefully measured upon a scale of inches and twelfths or lines, the several diameters were entered into a book of record, whence they were taken to make out his tabular statement.

Dr. Meigs presumed, that one hundred and fifty measurements might be taken as sufficient to give a valid mean of dimensions; yet it would be better to get such mean from one thousand or from one hundred thousand. He stated, that he should probably continue the habit which he now had, of measuring the crania of children born under his care as surgeon accoucheur.

It would be observed, he added, that he had taken only two diameters—the *occipito frontal,* and the *bi-parietal*—the former extending from the most projecting part of the forehead to the occiput, and the latter from one parietal protuberance to the other.

He had taken these only because he deemed them mainly important. From the transverse we have the perpendicular, as well as the smallest circumference, while the occipito-frontal diameter gives us also the horizontal circumference.

The occipito-frontal diameter is too variable to make it useful to get a mean of its length, as it may differ more than three inches from its normal length under circumstances of very protracted or violent labor.

Dr. Meigs remarked, that before presenting his tables to the Society it seemed proper to say something as to the diameters allowed by other persons, in order that they might be compared with his own results.

Observers.	*Measurements.*	
	Occip. Fr.	Bi-parietal.
Dr. Burns,	4.	3.3 to 3.6
Blundell,	4.	3.6
Ryan,	4.6	3.6
Ashwell,	4.	3.3
Ramsbotham,	4.6	3.6
Baudelocque,	4.3	3.4–6*
Maygier,	4.6	3.6
Madame Boiorn,	4.3–6	3.4–6
Charley,	4.	3.3–6
Gardien,	4.	3.6
Dugés,	4.	3.3
Velpeau,	4.	3.6
Moreau,	4.3	3.3–9

MM. Lallemand and Franc, editors of M. Dugés' Man. d'Obst. p. 68, say that they carefully measured the diameters in twenty-six children at birth, and found that for the

Occ. frontal, the mean was 4.1
Bi-parietal,   " 3.4

---

* N. B.—The Paris inch is 1.065977; the English inch being regarded as 1.0.

The maximum.. 4.3
Minimum 3.6

Dr. M. then presented a tabular statement of the measurements of one hundred and fifty heads of children, born *at term*, in the city of Philadelphia. It appeared from this:—

1. That the sum of his measurements of the occipito-frontal diameter was 729 inches, 7 lines; the mean 4 inches, 10 lines.

2. That the sum of his measurements of the bi-parietal diameter was 586 inches, 7 lines; the mean 3 inches, 11 lines.

3. That in 52 children, the occipito-frontal diameter was more than 5 inches, in 11 it was 5.1, in 8 it was 5.2, in 3 it was 5.3, in one 5.4, in one 5.6, in two 5.7, and in one 5.10.

4. That in 68 the bi-parietal diameter exceeded 4 inches, in 19 it was 4.1, in five 4.2, in six 4.3, in three 4.4, in one 4.5; in only one case was it less than 3.6: in that it was 3.4.

5. That the mean horizontal circumference of the head was 13.8, and the small or perpendicular circumference 11.5.

Dr. Meigs concluded with remarking, that perhaps in an ethnographical relation his measurement might possess a certain interest. It would be curious to preserve, from century to century, some accurate registers of the kind; and he ventured to hope, that upon many succeeding centennial sessions of the American Philosophical Society future members might be induced to present similar records.

Mr. Boyé read a communication "On the Conversion of Benzoic Acid into Hippuric. By James C. Booth and M. H. Boyé."

Wohler was the first to observe, that on administering benzoic acid to a dog, it was converted into an acid resembling hippuric acid. Ure afterwards found hippuric acid in the urine of a patient who had taken benzoic acid, and as he could discover no uric acid in the urine, he thought that this last was employed in the conversion of benzoic into hippuric, and recommended the former as a remedy for calculous concretions of uric acid. Keller, induced to undertake some experiments at the instigation of Wohler, found that by taking benzoic acid it reappeared in the urine as hippuric; that it continued as long as he continued the use of benzoic acid; that the urine contained both uric acid and urea; and that it could be inspissated without depositing hippuric acid, unless chlorohydric acid were added, showing that it was combined with a base.

The authors repeated Keller's experiments, and found that when the benzoic acid was taken after a meal, the uric acid which at first occurred in the urine did not diminish by the appearance of the hippuric, and that the uric still exhibited itself even during the subsequent continuance of the benzoic acid. The urine may be inspissated to a syrupy consistence without depositing hippuric acid or its compound: upon separating the syrup from the deposited salts, and shaking it with a mixture of 2 ether + 1 alcohol, urea is dissolved without any hippuric acid, while 1 ether + 1 alcohol dissolves both urea and the hippurate, but the former in so much larger proportion, that the hippurate is eventually extracted free from urea. The hippurate then proves to be a compound of the acid with, ammonia.

The authors draw the following conclusions from their experiments:—

1. That the formation of uric acid in healthy urine is not affected either in regard to its quantity, or to its external properties in general, by the introduction and transformation of benzoic acid into hippuric acid in the system.

2. The time required for the benzoic acid to pass through the system, and reappear as hippuric in the urine, is from twenty to forty minutes after its introduction with food into the stomach. Its occurrence continues for four to eight hours, but then ceases.

3. The quantity of hippuric acid obtained from urine is greater than that of the benzoic taken. In round numbers it may be stated to be one-third more.

4. Urea is not in combination with hippuric acid in the urine.

5. The base with which hippuric acid is combined, and which keeps it in solution, is ammonia.

Mr. Simeon Borden, of Massachusetts, presented a communication, entitled, "Comparison of the Dimensions of the Earth, obtained from Measurements made in the Survey of the State of Massachusetts, with accredited Mean Determinations."

In this paper Mr. Borden deduces the length of the meridional degree, the length of a degree perpendicular to the meridian, and the length of a degree of the parallel, for latitude 42° 21′ 30," from Professor Bessel's corrected elements of the magnitude and figure of the earth, derived from ten principal geodesical surveys, and compares these with the results of similar deductions from elements derived from the Massachusetts survey. He gives the results of his comparisons as follows:—

Equatorial radius of the earth, according to

Bessel,	20923597.14 feet, or	3962.803	miles.
Mass. survey,	20921935.54	″	3962.487	″
Difference,	1661.6	″	0.316	″

Polar semi-axis, according to

Bessel,	20853657.16 feet; or	3949.556	miles.
Mass. survey,	20850374.32	″	3948.935	″
Difference,	3472.16	″	0.621	″

Length of degree of meridian, for lat. 42° 21′ 30″, according to

Bessel,	364403.28	feet.
Mass. survey,	364356.0	″

Difference,   47.28   feet.

Length of degree of perpendicular, for lat. 42° 21′ 30″, according to

Bessel,	365740.21	feet.
Mass. survey,	365724.00	″

Difference,   16.21   feet.

Length of degree of parallel, for lat. 42° 21′ 30″, according to

Bessel,	270262.09	feet.
Mass. survey,	270250.10	″

Difference,   12.   feet.

Mr. Borden, after some remarks in explanation of the manner in which the position of different astronomical stations was determined by Mr. Paine, gives his reasons for rejecting three of his results, as irreconcilable with the rest, and probably referable to a deflection of the plumb line. He then proceeds to rededuce the latitude of the State House at Boston from a comparison of all Mr. Paine's admitted results. The indiscriminate mean of these fixes it at 42° 21′ 29″.16, which differs 0″.84 from that formerly arrived at by Mr. Borden. A comparison made by Mr. Boutelle of the same results, which gives a weight to each result proportioned to the number of observations on which it is founded, fixes it at 42° 21′ 28″.809. On the whole, Mr. Borden is inclined now to assume 42° 21′ 29″ as the latitude of the Boston State House, and to correct all the computed latitudes of Massachusetts accordingly, by deducting one second.

The paper is accompanied by an investigation of the ellipse, which contains the formula used by Mr. Borden in deducing the spheroidal elements compared in the earlier part of his communication.

Professor Bache gave an account of an instrument for determining the conducting power of bodies for heat.

The instrument depends upon the same principles with Fourier's thermoscope of contact. A conductor of heat of limited extent exposed to temperatures of different degrees at opposite sides, takes an intermediate temperature between the sources of heat. If the flow of heat on one side be cut off by interposing a substance, of which the conducting power is to be determined, the new temperature assumed by the conducting mass depends upon and measures the conducting power of the body interposed. In Fourier's thermoscope of contact an air thermometer was interposed between two vessels of water heated to different temperatures, and contact with the sheet or lamina of the substance whose conducting power was to be measured was obtained by using a cushion containing mercury. The air thermometer has always been found a very unmanageable instrument, and the mercury cushion secures only a partial contact.

Prof. Bache's thermoscope of contact consists of three compartments, formed by dividing a trough cut in a block of wood by thin partitions of appropriate material. Each compartment is filled in part with mercury, and a thermometer bulb in the center of each, and nearly filling it, indicates its temperature. A vessel containing ice fits closely to one end of the block in contact with one of the exterior compartments, and another to contain warm water, kept at nearly a constant temperature by a lamp placed beneath, fits closely to the other compartment.

If the resistance to the flow of heat is equal from each of the exterior compartments to the central one, the temperature in it will be the mean between the temperature, in the two exterior compartments, though in the apparatus as actually constructed by Prof. B. the relation is purposely made different from this. To measure the conducting power of a thin plate of wood or metal, which mercury does not wet, it is placed in a small groove in one of the exterior compartments, the mercury being in contact with it on both sides. The new temperature taken up by the central thermometer indicates the relation of the resistance to heat of the substances on the opposite sides of the central compartment, and this may be made to measure the conducting power of the interposed plate. Mercurial thermometers of which the degrees are minutely subdivided by verniers wore preferred to other modes of measuring the heat; and it was remarked, that as the actual equilibrium of temperature is in general only momentary, varying with the changes of the cold and warm vessels, it is of more importance to have a very rapid than a very minute reading of the scales.

Professor Bache explained how this instrument may be applied to measure the conducting power of substances used for clothing, imperfectly investigated by Rumford,—of various metals, of powders of different kinds, and of fibrous substances; and remarked that a simple modification would render it applicable to the measurement of the conducting power of liquids. Though the principle of contact in the instrument and the other details are essentially different from those of Fourier's instrument, Prof. Bache prefers, in order not to introduce a new name into science, to retain for it the name given by Fourier to his instrument intended for the same purpose.

Professor Bache remarked, that a different view might be taken of the theory of a thermoscope of contact, which would lead to a corresponding modification of the instrument. If a trough containing mercury be heated at one end, the heat will be propagated according to a well-known law. If a partition be placed in the trough, a new distribution of temperature will take place, which when compared with the former will give the conducting power of the substance obstructing the flow of heat.

Professor Henry read a letter, addressed to him by Captain Thomas Lavender, of Princeton, N. J., giving his recollections of some water spouts observed by him. The following are extracts from this letter:—

The most remarkable of these was one which passed directly across our ship, the *Huntress*, of New York, bound for China, in June 1822; we were in the latitude of between three or four degrees north, and longitude of from twenty-three to twenty-five west, which, at that season of the year, was near the northern limit of the southeast trade wind. The ship was sailing on a south-westerly course, with the wind from south to south south-east; our attention was first directed to the spout by a loud roaring noise, and we saw at the distance of about three or four hundred yards, directly to windward, a circular space on the water of about twenty feet in diameter, within which the surface was violently agitated, and appeared to be raised to the height of a foot into a perfectly white foam. We saw no appearance of a cloud above, and no descending column, but this might have been because we were too near the spout; since I have observed, on another occasion, that the regular outline usually described vanishes when the object approaches very near the observer. From its direction when first observed, and our rate of sailing,—about four miles per hour,—it was at first thought that it would pass us two or three hundred yards astern, if it moved in the direction of the wind; but instead of this, it appeared to keep on our beam, and its velocity to increase as it neared the ship. Fortunately, all hands were on deck at the time; and when it became evident that it would strike the ship, the order was given to lower the sails. Simultaneously with the act of lowering them, we felt the effects of the spout. The ship was thrown over in an angle of fifty degrees with the horizon, as if by a wind directly from the spout. It passed across the ship in from ten to fifteen seconds, during which time we were wet with a hard shower of rain composed of very large drops of water. As the spout was leaving the ship on the other side, all the sails throughout the whole length of the vessel, say one hundred and sixty feet, were thrown aback by a wind from the spout in an opposite direction, and the ship was inclined into an opposite angle of about the same number of degrees as before.

The action of the spout appeared to be very much circumscribed, since we did not feel the wind from it until the agitation on the water approached within twenty yards of the side of the vessel. The effect appeared to continue longer as it was leaving us than it did when it was approaching. After leaving the ship, it was observed to move in the general direction of the wind; but it appeared to have lost considerable of its force, and did not make as great an agitation on the water as before.

The time was about sundown; the day had been squally until within an hour or so, when the weather cleared off with passing clouds. No whirling motion was observed; the continued noise and the circular spot on the water were the only effects noted until we were struck by the wind.

Three years previous, while in the Straits of Gaspar at the entrance of the China Sea, we saw fourteen water spouts in the course of an hour. One of these, and the largest I ever saw, was within two hundred yards of the ship. Its action on the water covered a circular space, I should say, of one hundred feet in diameter; it was furious, and threw up a cone of spray quite dense, of from three to four feet in height, something like a large sugar loaf. It commenced and exhausted itself on the same spot: no wind was felt from it, and the ship was becalmed the whole time, although a heavy rain was falling on us, which was not, however, of the same, intensity continually: when the spout subsided, the rain ceased. We fired a number of shots from a six-pounder directly through it, without any apparent effect. The interest of the scene was much increased by the presence of another ship, which was also occupied in the endeavor to dissipate other spouts by the same means. Indeed we both became so much engaged in this work, that each ship was endangered by the balls from the other.

On another occasion, off the Cape of Good Hope, we saw a spout fully formed at the distance of about a mile. It presented the usual appearance of a cloud above, with a spiral waving column reaching down to the water, and agitated by a continued motion like that exhibited by a piece of twisted glass turned on its axis before the eye. Its motion toward us being slow, we secured all the sails, and watched it with a telescope. When it was at the distance of about one hundred yards, I could see distinctly into the center of the base; which was perfectly smooth: not a drop appeared to disturb the surface; while without the circle, and all around its circumference, the sea was in bubbling commotion, with large drops of rain falling on it, some of which reached the ship. No wind was felt from it. The base, I should judge, was about fifty feet in diameter. As it approached us, the distinctness of the outline of the spout became less apparent, and a diffuse mistiness appeared to occupy the place. Its intensity began to decline about the time the telescope was directed toward it, but it continued for some time after in tolerably violent action. It passed us at the distance of about thirty yards astern, where its action ceased in plain sight. The time of day was about an hour before noon.

The foregoing are all the spouts which I recollect, as being very near to us.

Off the Isle of France, one passed apparently across the ship's course at the distance of about a mile astern; the wind blowing a gale from the N. E.; ship steering west. Its direction was not the same as that of the wind, but apparently nearly at right angles to it. Its motion was very rapid; I should think at the rate of twelve miles per hour; and as it crossed our course, it presented the appearance of an immense black snake, with its head in a cloud, its body dragging far behind, and the tip of its tail touching the surface of the ocean.

Mr. Richard C. Taylor made some observations respecting the appearance of certain water spouts which he had observed on the Bahama banks.

During several days in the month of June 1836, on a voyage from the port of Guiana to Havana, along the north coast of Cuba, the vessel on board of which Mr. Taylor was, was becalmed several days, both on the Grand Bank, and in the main Bahama channel.

During that time, several water spouts appeared around the ship, at from two to four miles distance. Being at the time little conversant with such phenomena, and perceiving that they presented a variety of aspects, Mr. T. sketched, at the moment of their occurrence, three of the most characteristic; each of them occurring on a different day. These drawings he presented to the meeting.

The first of the water spouts occurred about noon, on the 10th of June. By observation, at this hour, the sun's corrected altitude was 89° 44', and the latitude 23° 16'. The temperature of the air at the moment of observing the water spout was 84°—and the temperature of the water of the sea was 82°. Having never until that moment observed a water spout, Mr. T. was somewhat puzzled at the first glance, to know the true character of the object that thus attracted his attention. No inverted cone or funnel-shaped mass was remarked. But beneath a very black and not very extensive cloud, two masses or curved columns of misty vapor arose froth the surface of the sea, resembling in appearance the smoke of a steam-boat, or rather the two bodies of smoke were separated at the surface by a space which seemed scarcely wider than the breadth of a steamer would appear to be at that distance. The vapor or water of these bodies became momentarily more opaque, and evidently more agitated. What circumstances attended the further progress and accumulation of this disturbed double column, does not appear from Mr. T.'s notes. He perceived no descending column from the overhanging cloud; and it seemed to have suddenly vanished.

The following extracts from his diary were presented by Mr. Taylor:—

11th June.—In only four and a half fathoms water on the bank this morning; the bottom very white, with the coral mud, and the groups of Fuci growing upon it, very visible. The thermometer 88° in the forenoon at eleven: at twelve, 90°. Altitude of the sun, 89° 41′, and the latitude, 23° 15′. In deeper water this afternoon, as shown by its dark indigo color. Storms and thunder all round us this day, but seemed to be at a distance, fortunately.

Another water spout occurred to-day. The same form appeared below, as was seen in that of yesterday, viz., two columns curving outwards. A heavy lowering black cloud overhung the sea; and a dark, conical body, tapering downwards, apparently descended half from the cloud, to meet the ascending columns beneath.

12th June.—Perfectly calm in our position: some rain reached us for a short time: very hot in the middle of this day. Our vessel appeared rather to drift astern, as the fishing lines which had been thrown astern always kept ahead of us during two days. Several water spouts appeared around us. Some of them were bent, and some of them straight. In all cases, the double columns, resembling at first the smoke of fires, commenced below, before any cone passed down from the black clouds that hovered above, to meet them.

Professor Henry made some remarks on the subject of these communications. He thought that sufficient attention had not been paid to the influence of electricity in the production of storms and in the phenomena of meteorology generally. He adverted to the views of Dr. Hare and to those of M. Peltier, who has given a form to the electrical theory, which in Prof. Henry's opinion is exceedingly ingenious and worthy of attention.

Invitations were presented from Professor Kendall, to inspect the Astronomical, Observatory at the High School, and from Dr. Charles Frederick Beck, to examine the compound microscope of Powell and Lealand, belonging to him, after the adjournment.

Dr. Patterson, Director of the Mint of the United States, invited the members and visitors to witness the operations of the Mint, before the meeting of to-morrow morning.

SPECIAL MEETING.

Sixth Session, 29th May, half past 7 o'clock, P. M.

DR. PATTERSON, Vice-President, in the Chair.

Professor Frazer presented a written communication, entitled, "Transformation of the Series $S = ax + bx^2 + cx^3$, &c., by Professor Theodore Strong, of Rutgers College, New Brunswick, N. J."

It is evident that $x = \dfrac{1}{\dfrac{1}{x}} = \dfrac{1}{\dfrac{1}{x} - v + v}$ and if we put $y = \dfrac{1 - vx}{x}$

or $y^{-1} = \dfrac{x}{1 - vx}(1)$, we get $x = (y + v)^{-1} = y^{-1} - vy^{-2} + v^2 + v^2y^{-3} - v^3y^{-4}$

$+$ &c., $x^2 = y^{-2} - 2vy^{-3} + 3v^2y^{-4} - 4v^3y^{-5} +$ &c., $x^3 = y^{-3} - 3vy^{-4} + 6v^2y^{-5}$

$- 10v^3y^{-6} +$ &c., $x^4 = y^{-4} - 4vy^{-5} + 10v^2y^{-6} -$ &c., and so on for x^5, x^6, & c. Hence if we substitute of x, x^2, x^3, &c. in the given series, it becomes $S = ay^{-1} + [- av + b]\ y^{-2} + [av^2 - 2bv + c]\ y^{-3} + [- av^3 + 3bv^2$

$- 3cv + d]\ y^{-4} + [av^4 - 4bv^3 + 6cv^2 - 4dv + e]\ y^{-5} +$ &c.; or substituting

the value of y^{-1} from (1), we get $S = a\dfrac{x}{1 - vx} + [- av + b]$

$\dfrac{x^2}{(1 - xv)^2} + [av^2 - 2bv + c]\ \dfrac{x^3}{(1 - vx)^3} +$ &c. (2), whose law of continuation is evident; and it is manifest that if we expand the functions

$\dfrac{x}{1 - vx}$, $\dfrac{x^2}{(1 - vx)^2}$, &c., according to the ascending powers of v, the terms which involve v will mutually destroy each other, and (2) will be reduced to the given series; so that v in (2) is arbitrary, as it ought to be. If in the given series we change S and x into $- S, - x$, it becomes $S = ax - bx^2 + cx^3 - dx^4 + $ &c. (3), whose transformation by (2) is $S = a \dfrac{x}{1 + vx} - [- av + b] \dfrac{x^2}{(1 + vx)^2} + [av^2 - 2bv + c]$

$\dfrac{x^3}{(1 + vx)^3} - $ &c. (4). If we put $v = 1$, and use Δ for the characteristic of finite differences, then (since $\Delta a = - a + b$, $\Delta^2 a = a - 2b + c$, $\Delta^3 a$ $= - a + 3b - 3c + d$, and so on,) (2) and (4) become $S =$

$a \dfrac{x}{1 - x} + \Delta a \dfrac{x^2}{(1 - x)^2} + \Delta^2 a \dfrac{x^3}{(1 - x)^3} + \Delta^3 a \dfrac{x^4}{(1 - x)^4} + $ &c. (5). $S =$

$a \dfrac{x}{1 + x} - \Delta a \dfrac{x^2}{(1 + x)^2} + \Delta^2 a \dfrac{x^3}{(1 + x)^3} - $ &c. (6). (5) is given by Euler at page 283 of his Inst. Cal. Dif. If a, b, c, &c., have constant differences, (5) and (6) will give S the generating function of the given series, and (3). If $a = b = c = d = $ &c. (2) and (4) become

$$S = a \frac{x}{1 - vx} + a \, (1 - v) \frac{x^2}{(1 - vx)^2} + a \, (1 - v)^2 \frac{x^3}{(1 - vx)^3} + \text{&c.,}$$

$$S = a \frac{x}{1 + vx} + a \, (v - 1) \frac{x^2}{(1 + vx)^2} + a \, (v - 1)^2 \frac{x^3}{(1 + vx)^3} + \text{&c.,}$$

whose generating functions are $S = \dfrac{ax}{1 - x}$, $S = \dfrac{ax}{1 + x}$ v vanishing from the generating functions as it ought to do.

By the aid of v we can change the same series into an indefinite number of forms; for example, if in the last of the above series we put $v = 1 + x$, it will be changed to $S = a \dfrac{x}{1 + x + x^2} + \dfrac{ax^3}{(1 + x + x^2)^2}$

$+ \dfrac{ax^5}{(1 + x + x^2)^3} + $ &c., which has the same generating function as (3), although its form is very different.

It may be observed that (2) and (4) will be exhibited under a more general form by changing v into $\dfrac{v}{w}$, and they will by a slight reduction become $S = a \dfrac{wx}{w - vx} + [- awv + bw^2] \dfrac{x^2}{(w - vx)^2} + [awv^2 - 2bw^2v + cw^3] \dfrac{x^3}{(w - vx)^3} + $ &c. (2'); $S = a \dfrac{wx}{w + vx} - [- awv + bw^2] \dfrac{x^2}{(w + vx)^2} + awv^2 - 2bw^2v + cw^3] \dfrac{x^3}{(w + vx)^3} - $ &c. (4).

To show the use of our formulæ, we shall take the well-known series, log. $(1 + x) = x - \dfrac{x^2}{2} + \dfrac{x^3}{3} - \dfrac{x^4}{4} + \dfrac{x^5}{5} - $ &c., the logarithms being hyperbolic. Comparing this series with (3), we have $a = 1$, $b = {}^1/_2$, $c, = {}^1/_3$, $d = {}^1/_4$, &c.; hence by (4'), if we assume $w = 2$, $v = 1$, we get log. $(1 + x) = S = 2 \left[\dfrac{x}{2 + x} + \dfrac{1}{3} \dfrac{x^3}{(2 + x)^3} + \dfrac{1}{5} \dfrac{x^5}{(2 + x)^5} + \text{&c.} \right]$; and if we put $x = \dfrac{z}{n}$ and use logarithms whose modulus is M, we get log.

$(n + z) = $ log. $n + 2M \left[\dfrac{z}{2n + z} + \dfrac{1}{3} \left(\dfrac{z}{2n + z} \right)^3 + \dfrac{1}{5} \left(\dfrac{z}{2n + z} \right)^5 + \text{&c.} \right]$ a well-known result; also if we put $w = v$ we get by (4^1), log. $(1 + x)$

$= \dfrac{x}{1 + x} + \dfrac{1}{2} \left(\dfrac{x}{1 + x} \right)^2 + \dfrac{1}{3} \left(\dfrac{x}{1 + x} \right)^3 + \dfrac{1}{4} \left(\dfrac{x}{1 + x} \right)^4 + $ &c.; if this is multiplied by M, we shall have log. $(1 + x)$ in the system whose modulus is M, and if we put $\dfrac{x}{v}$ for x, it becomes log. $(v + x) = $ log.

$v + M \left[\dfrac{x}{v + x} + \dfrac{1}{2} \dfrac{x^2}{(v + x)^2} + \dfrac{1}{3} \dfrac{x^3}{(v + x)^3} + \dfrac{1}{4} \dfrac{x^4}{(v + x)^4} + \text{&c.} \right]$. In a similar manner, by giving different values to v and w, we may obtain an indefinite number of forms for log. $(1 + x)$; but if we would compute the numerical value of log. $(1 + x)$ for any particular value of x, we must take a form whose successive terms decrease, and the more rapidly the better; if the terms should (instead of decreasing) increase, no part of the series can with safety be assumed for the numerical value of log. $(1 + x)$: and similar remarks are applicable in all cases where it is proposed to calculate the numerical value of any generating function by means of series, or, as it is often very improperly said, to sum the series.

Mr. J. N. Nicollet, of Washington City, exhibited his original map of the North Western Territory of the United States, made from personal observations, and read an account of his geographical exploration of the sources of the Mississippi.

Mr. Nicollet left St. Peters on the 26th July, 1836, and the Falls of St. Anthony on the 29th. Having arrived at Crow Wing river, 189 miles above St. Peters, he left the line which had been before explored by Major Pike and other Americans, and directed his course by the Gayank or Gall river, and other streams, to Leech lake. Having succeeded, by the instrumentality of a missionary whom he found there, in conciliating the jealousies of the natives, he again set out in a bark canoe, with three attendants. Crossing several small lakes to the lake Kabekonang, he ascended the river of the same name. This stream flows in a narrow and deep valley, and is said not to freeze before January, and when frozen, not to thaw before July. Mr. N. found in August, that its waters had a temperature of only 54°, whilst that of the lakes and rivers, which he had noted the preceding days, was between 60° and 70°. It is protected by dense overhanging forests, and has but one small tributary creek, being abundantly fed by springs. It is in latitude 47° 16′, the mean annual temperature being 43°.

From the sources of the Kabekonang by a portage of five miles, he reached the river La Place, which he ascended to the vicinity of Assawa lake, and thence by a painful and difficult portage of six miles he passed to lake Itasca.

This lake, the *Omoshkos Sagaigon* of the Chippeways, the *Lac à la Biche* of the French, and *Elk lake* of the British, has been regarded as the fountain of the Mississippi. It is however supplied by five creeks, which are formed from innumerable streamlets, oozing out of the clay beds at the bases of the "Hauteurs des Terres," or land heights. These elevations consist of accumulated sand, gravel, and clay, intermixed with erratic fragments. They are commonly flat at top, and vary in height from eighty-five to one hundred feet above the level of the surrounding waters. They are covered with thick forests, in which the coniferous plants predominate. South of Itasca lake, they form a semicircular region with a boggy bottom, extending to the S. W., a distance of several miles. Thence they ascend to the N. W. and N., and then stretching to the N. E. and E. through the zone between 47° and 48° of latitude, make the dividing ridge between the waters that empty into Hudson's Bay, and those which discharge themselves into the Gulf of Mexico.

The principal group of the Hauteurs des Terres is subdivided into several ramifications, varying in extent, elevation, and course, so as to determine the hydrographical basin of all the innumerable lakes and rivers that characterize that region. One of them extends in a southerly direction, under the name *of Coteau du Grand Bois*, and it is this which separates the Mississippi streams from those of the Red river of the north.

The waters supplied by the north flank of these heights,—still on the south side of Itasca,—give origin to the five creeks which flow into that lake. These waters Mr. N. considers to be the utmost sources of the Mississippi. Those that flow from the southern side of the same heights, and empty themselves into Elbow lake, are the extreme sources of the Red river of the north; so that the remote feeders of Hudson's Bay and the Gulf of Mexico approximate closely to each other.

Of the five creeks that empty into Itasca lake, one enters its eastern bay; to this Mr. N. has given the name of La Place, and a lake through which it passes he has called after the translator of the Mécanique Celéste, lake Bowditch. But he regards as the infant Mississippi, one of the four which enter the western bay of Itasca, equal in length with the La Place, and more abundant in its supply of water.

On the 29th of August, 1836, this stream, at its entrance into Itasca lake, was from fifteen to twenty feet wide, with a depth of water of from two to three feet. Mr. N. explored it for some miles to its highest streamlets, and determined the mean temperature of the region by plunging a thermometer in the mud from which the springs issue. He found it, on repeated trials, to be between 43° 5′ and 44° 2′ F.; that of the air being 63° and 70° at the times of observation.

These first waters of the Mississippi unite at a short distance from the hills in which they originate, and form a small lake, from which the rivulet passes with a breadth of a foot and a half, and a depth of a foot. Soon however uniting with other streamlets, this supplies a second minor lake, the temperature of which was ascertained to be 48°; and issuing from it with increased velocity and volume, enters a third lake of somewhat larger proportions, from which, by a channel of two or three miles, it passes into lake Itasca.

The only island in Itasca lake is but about 222 yards long. The Mississippi, on issuing from the lake, is sixteen feet wide, having a depth of fourteen inches, perfectly transparent, with a swift current. The temperature of the water at seven o'clock in the morning was 62°, while that of the air was 56°. After an hour's descent, Mr. N. found that the breadth had increased to twenty-five feet, and the depth to three feet.

The communication of Mr. Nicollet proceeded to describe the geographical character of the entire upper Mississippi, as observed by him, its geology, capacities for production, facilities of communication and defense, &c. &c. It is intended to form a part of the elaborate report, now nearly completed by him, of his explorations in that region; and it is therefore less to be regretted that its condensed character precludes a satisfactory abstract of its diversified contents.

Mr. Lea made an oral communication "On Coprolites."

Mr. Lea said, that in this communication he did not profess to introduce any important new facts in connection with these interesting petrifactions. He was, however, desirous of showing their geological importance, and of exhibiting his collection of foreign specimens and casts; hoping to draw the attention of his fellow-laborers in this country to a subject but little attended to here.

The introductory part of Mr. L.'s remarks defended the science of geology from the allegation, that observations made in the stratified masses of the earth are of course inconclusive; and he instanced some facts, which he conceived to be as conclusive and precise as the certainties of mathematics. He observed that, "if we see in a homogeneous mountain mass the impress only of a fractured portion of a shell or other organic matter, we may be enabled to assign to that mass its proper situation in the great series of formations which compose the crust of our planet. This however does not strike the man of science with any thing like the same wonder as the gradual but certain decadency of the animated beings which inhabited the successive strata."

Mr. Lea exhibited various large drawings of fossilised ejectamenta, and commented on the spiral character of the intestinal passages which must have produced them; and he produced for the sake of comparison specimens of fæces from the living alligator, taken from the banks of the Savannah river, which bore a strong resemblance to one of the foreign coprolites on the table. He observed that coprolitic remains are of rare occurrence in the United States. They had been noticed by Dr. De Kay in the green-sand formations of Monmouth, N. J., by Professor H. D. Rogers at Crosswick's Creek, N. J., and by Dr. Morton in the green-sand of Virginia. Their rarity here, he attributed to the absence in this country of those formations, in which they most abound in England, viz. the strata included in the oolitic groupe.

Mr. L. concluded his communication by calling the attention of the Society to a splendid specimen of *Pentacrinites Briareus*, belonging to his cabinet, from the Lias of England. He described the animal and its functions with the assistance of an enlarged drawing prepared by himself for the occasion.

Dr. B. H. Coates read a Memoir "On the Effects of Secluded and Gloomy Imprisonment on Individuals of the African Variety of Mankind in the Production of Disease."

This was founded partly on statements of the mortality in the Eastern Penitentiary of Pennsylvania, since its organization, thirteen years ago; partly on the assigned causes of death during four years of that period; and partly on the physiological characters of the African. The statements contained in a memoir he has transmitted to a German Journal, were recapitulated and rendered complete by the addition of the year 1841, which he had previously not included from a defect in the materials.

Dr. C. disavowed any intention of censuring the practice or assailing the mode of punishment by separate confinement, unless in the case of colored persons; the Eastern Penitentiary being used as an example, because he had been during the above whole period an official visitor of that institution, and because it was the one from which it was most easy to obtain the requisite information. Indeed, it appeared that the mortality of the white persons confined in the prison was less than that of the whites in the City and Liberties of Philadelphia. He had not been able to procure in time the population within the bills of mortality, owing to the manner in which the census is published; but was obliged to make use of the reports collected and analyzed by Dr. G. Emerson for the ten preceding years, extending from 1820 to 1830. Comparing these with the reports from the penitentiary, the relative mortality stood thus:

	Per cent.
White mortality in penitentiary, average of 13 years, ending January 9, 1843,	2.03
White mortality in city and suburbs, average of 10 years, ending Jan. 1, 1831,	2.422
Colored ditto, city and suburbs,	4.752
Colored ditto, penitentiary, 13 years,	7.03

Assuming Dr. Emerson's mortality of whites as unity, it was—

| Whites in penitentiary, | 0.838 |

Whites in city and suburbs,.. 1.
Colored in city and suburbs,.. 1.962
Colored in penitentiary, ... 2.903

The mortality of the colored in the penitentiary is found to bear to that of the whites, the proportion of 7.03 to 2.03, or of 346 to 100.

Tables were given, further illustrated by curved lines drawn through ordinates, to illustrate the succession and fluctuations of these rates of mortality. The population of the prison had not become steady till 1836; which renders the average of the first six years more variable and of less value; but after the expiration of that time, a great excess not only in the proportion of deaths, but in the fluctuation of that proportion was visible among the colored population of the prison. The effect of epidemics was insufficient to account for the discrepancy.

It is necessary here to bear in mind that the convicts in the penitentiary are not, as has been sometimes supposed, the most wretched and most exposed to hardships of our population. The most miserable, and in particular the most miserable blacks, seldom commit the higher crimes which render them liable to the larger periods of confinement for which the penitentiary is intended; but are generally either convicted of lighter offenses or committed for vagrancy. In both cases they are sent to the Moyamensing prison. Besides; the immediate effects of drunkenness, recent colds and violence, have generally had time to subside before the prisoners are sent to the penitentiary; as, prior to this, they must undergo their trials, and remain committed in the other prison, if not bailed.

It is evident, that a comparison on terms of perfect equality cannot be made between the white mortality in the prison and that in the city. Of causes tending to diminish the proportion of deaths in the prison, one of the most important is that the convicts are generally persons in the prime of life, and that the prison is exempted from the heavy mortality of infancy, and from that of old age. On the other hand, there must be admitted to occur among the convicts, a large proportion of individuals who have injured their constitutions by a vicious mode of life. With this proviso, there is a convenience in placing together these four ratios in a common view, as follows:

	Per cent.
White mortality in penitentiary, 13 years,	2.03
" " city and suburbs, 10 years,	2.422
Colored " " " "	4.752
" " penitentiary, 13 years,	7.03

If Dr. Emerson's average of white deaths in the city and suburbs be assumed as unity, these numbers will then be represented as follows :—

	Per cent.
Whites in penitentiary,	.838
Whites in city and suburbs,	1.
Colored in city and suburbs,	1.962
Colored in penitentiary,	2.903

Dr. Coates did not doubt the adequacy of these statements to establish the fact, that there exists an immense discrepancy in the effect of imprisonment between the colored people and the whites; and that there is an essential difference in this, as in so many other respects, between the two races. The most prolonged and minute inquiry has failed to discover any difference in the treatment of these two classes in the prison; unless it be, that, from the dislike of cold, the colored convicts frequently deprive themselves of a portion of their ventilation.

Assuming, then, as the result of this evidence, that there does exist the difference in question between the two races, Dr. Coates passed to the physiological observations which tend to confirm, and at the same time explain it.

The negro, or even the mulatto, is a very different person, in his physical and psychical conformation, from that one who may be presumed to have been held in view in our legislation, the white Anglo-Saxon, Celt, or German. His ancestry and the prototype of his race belong to the torrid zone; and even the mixed progeny suffer severely and mortally by our cold. Cheerful, merry, lounging and careless, the Ethiopian American deeply enjoys the sun and light; delights in the open air; and is, as a general rule, constitutionally free from that deep, thoughtful anxiety for the future, so conspicuous in his paler neighbor. The face of heaven seems to him necessary to his existence; and though long confinement is in his case less productive of gloomy remorse, it is far more depressing to his vitality.

The morbid effects of this have been already mentioned, in the production of pulmonary consumption and scrofula; more than 88 per cent of the deaths being from chronic affections of the lungs and from the last named disorder. The moral consequences are in an equivalent degree depressing to the mind. It is not by remorse and anguish that he is affected, so much as by intellectual and moral weakness and decay; and gloomy confinement becomes thus to him, mentally as well as physically, a nearer approach to the punishment of death.

Dr. C. controverted the opinion that the effect of separate imprisonment has been to produce insanity; though a humane and strict analysis, he said, has shown many to have been affected both with insanity and with imbecility at the times when they committed the offenses for which they were sentenced. The effect, scarcely perceptible upon the whites, has been upon the unfortunate colored prisoners to produce, not mania, but weakness of mind; dementia, instead of deranged excitement.

Recurring to the evidence furnished by the official reports from the Penitentiary, Dr. Coates remarked, that in the reports for 1837, 1838, 1839, and 1842, a detail is given of the mortal diseases and their immediate causes in forty-three cases of colored persons. These are as follows:—

Consumption and chronic inflammation of the lungs; 1837,
    6 cases; 1838, 12; 1839, 1; 1842, 1,.. 20
Scrofula of the chest, 1838, 1, ... 1
Chronic pleurisy, 1838, 2 cases; also affected with chronic
    inflammation of the stomach, or with that of the bladder,
    and with paralysis: 1839, 5; of which 1 was cut off by
    brain fever: 1842, 1, ... 8
Scrofula, of other parts than the chest, 1837, 2 cases; 1838, 4,
    including affections of peritoneum, bowels, and knee joint;
    1839, 2, including peritoneum and hip joint; and 1842, 2, ... 10
Typhus fever, 1837,... 1
Remittent fever, 1837, ... 1

Asthenia, 1842, ... 1
Tetanus, from a burn, 1842, ... 1

                                               Total,   43
                                               ————

Vicious habits are enumerated as causes of fourteen of the cases; and in three of them, they are the only cause assigned. Four are ascribed to previous syphilis; and in one, no other cause is recorded.

Of nineteen deaths of white prisoners, during the same years, the diseases were as follows:—

Consumption, 1837, 5 cases; 1838, 3 cases, ... 8
Pulmonary and hip disease, 1842, ... 1
Brain fever, reported as owing to scrofula and disorganized
    lungs, 1837, ... 1
Syphilitic chronic pleurisy, 1839, ... 1
Scrofula, ... 0
Syphilis, 1837, ... 1
Chronic bowel complaints, 1838, 1; 1842, 1, ... 2
Small pox, 1838, ... 2
Asthenic brain fever, 1839, ... 1
Stone, 1838, ... 1
Diseased arteries and enlarged heart, 1842, ... 1

                                               Total,   19
                                               ————

Vicious habits are assigned as a cause in four cases; and in two are the only cause named. Vice before admission is represented as a cause in four cases; and in two of them is the only one named.

Of forty-three deaths of colored convicts, twenty-nine are ascribed to chronic diseases of the lungs, and to affections of an adjacent structure, in which these are extremely liable to produce such diseased changes; and ten to scrofula of other parts than the lungs; leaving only four for all other affections; and these four were produced by typhus fever, remittent fever, asthenia, and tetanus. Of nineteen deaths among white persons, on the other hand, the causes are found in chronic diseases of the lungs and their appendages, for eleven cases; scrofula, none; chronic bowel complaint, two; small pox, two; and four others are severally attributed to syphilis, asthenic brain fever, stone, and diseased arteries, with enlarged heart. Reduced to percentages, these proportions would read as follows:—

Colored.

Diseases of chest, exclusive of heart and arteries, ... 65.12
Scrofula, ... 23.25
All other diseases, ... 11.63
                                                                                         ———
                                                            Total,   100.

White.

Diseases of chest, ... 57.89
Scrofula, ... 0.
All other diseases, including bowel complaints,
     10.53, and small pox, 10.53, ... 42.11
                                                                                         ———
                                                            Total,   100.

Mr. Richard C. Taylor read a paper "On Fossil Arborescent Ferns of the family of Sigillaria, occurring in the roof and floor of a Coal Seam in Dauphin County, Pennsylvania."

The author observed, that writers on fossil botany had almost entirely restricted their illustrations to portions or fragments only of the larger coal plants, for the purpose of displaying the characteristic cicatrices and other marks upon their stems, and the structure and arrangement of their leaves. Little has as yet been done toward an actual investigation of the magnitude to which this magnificent flora attains, in the stratified beds of our carboniferous regions. The opportunities which naturalists possess for observing the complete development of the larger plants or trees are rare. M. A. Brogniart cites only one instance where he had measured the stem of a sigillaria, which was forty feet in length and one foot in its greatest diameter.

Upon the geologists, and more especially upon those who direct explorations of mines, the charge mainly devolves of observing and recording the phenomena of fossil vegetation upon the large scale.

Recent opportunities afforded the writer of this paper the means of examining some of the largest specimens of sigillaria of which we possess any record. They occur in a coal seam in the western part of the Schuylkill coal field, on the upper and lower walls of a gallery conducted several hundred feet along a coal seam.

The *floor* of this coal, as usual in Pennsylvania, consists of indurated clay; the "bottom slate" of the colliers. Upon the surface of this are impressed innumerable well-preserved specimens of several species of sigillaria; a class of plants which is admitted into the family of tree ferns. More than a hundred of these are exhibited in the drawing illustrating this paper. Of this series, very few of the trunks of these trees are seen here of a less diameter than two feet; many are three feet; several are four and four and a half feet, and one, at least, is five feet wide. In no instance has the area of excavation been extensive enough to exhibit either of the extremities of these enormous stems, notwithstanding that many of them are laid bare for thirty, forty, and fifty feet of their length, without much apparent diminution or tapering upwards.

The *roof* or north wall is of siliceous conglomerate; between which and the coal is an extraordinary assemblage of curving trunks of arborescent ferns of the family of sigillaria. Some of them appear to be dichotomous, and to possess the characters of *S. elegans.* Such is the scale of these plants, that the extent of cleared space was, as in the floor, inadequate to elucidate the entire development of their gigantic stems. One example, although laid bare more than fifty feet, shows no signs of either termination, and looks as if it might extend thirty or more feet further. Another exhibits sixty-five feet in length of a flexuous stem, which apparently extended at least thirty feet beyond. A third, the most interesting of the group, shows at its base what obscurely seems to be the root. Near the base, the stem is about two and a half feet in diameter or breadth; forty feet up the trunk, it measures two feet broad, and continues in about this rate of diminution. Seventy feet of this specimen is above the floor of the gallery; it was traced several feet further below the floor, and in all was perhaps eighty to one hundred feet; but of this, and of the character of that superior termination, we have no present knowledge.

Mr. Taylor applied to this interesting illustration of the ancient flora, Mr. Logan's views as to the universal prevalence of the plant stigmaria in the argillaceous coal floors of coal seams, and its absence in the roofs. In the present instance, where a surface of seven or eight thousand square feet has been denuded, stigmaria are but rare. Only two well-defined specimens have been observed. One of these is seen in the roof above the coal; the other in the floor below the coal.

Six species of fossil plants were observed in the roof, and seventeen species in the floor.

An invitation to the members and visitors was presented from the Mercantile Library Company, to be present at a discourse to be delivered by Dr. Gouverneur Emerson at Laurel Hill Cemetery, on Thursday afternoon, upon the completion of the monument erected by the Company over the remains of Thomas Godfrey, the inventor of the Mariners' Quadrant.

SPECIAL MEETING.

Seventh Session, 30th May, 10 o'clock, A. M.

DR. PATTERSON, Vice-President, in the Chair.

A letter was read from Captain Ericsson, dated New York, 27th May, on the subject of the Centennial Celebration.

Professor Kendall read a communication, entitled, "On the Instruments of the Astronomical Observatory of the United States' Military Academy, West Point,—and on the Observations made upon the Comet of February, 1843,—by Prof. William H. C. Bartlett, of the U. S. Military Academy."

The first part of this paper gives an account of the new building erected for the accommodation of the library and philosophical apparatus of the Military Academy. It is built principally of granite, and stands at the S. E. angle of the plain, about 160 feet above the level of the river; having an uninterrupted view to the south of about eight miles, and to the north of about four miles. The main cell is one hundred and twenty feet long, and sixty feet broad. It is divided into two equal parts by a partition wall. The western division, with the tower, being appropriated to Natural and Experimental Philosophy, for purposes of instruction; and the eastern division, which is in one entire room, to the library. The astronomical instruments are in three towers, prepared expressly for them. Within these towers are masses of masonry, which rise from a bed of coarse gravel, twelve feet below the natural surface of the ground, to a height necessary to free the view from all obstructions. These columns are perfectly insulated from the foundation to the top, where the instruments rest. Immediately to the south of the central tower, is a fourth insulated column, extending nearly to the ridge of the roof, with no covering: this is intended for observations with the portable instruments. The central tower is surmounted by a revolving dome, twenty-seven feet in horizontal diameter, and about seventeen feet high from the springs. It has five window openings near the curb, and an observing slit, two feet wide, extending from a point forty-eight inches above the floor, to nearly two feet on the opposite side of the zenith: this has a range of shutters which are worked by levers, independently of each other. The whole dome rests on six twenty-four-pound cannon balls, which turn between two cast iron annular grooves. The dome is moved by means of a rack, attached to its base, and a stationary pinion, which is turned by a hand wheel, seven feet in diameter. The flank towers are furnished with meridian observing slits, about twenty inches in the clear, which afford an uninterrupted view of the celestial meridian.

The building itself, which is of the Elizabethan style of Gothic architecture, was designed by Major Delafield, the superintendent of the institution.

The second part contains a minute description of the equatorial, the only fixed instrument now in place. It is by Mr. Thomas Grubb of Dublin. Mr. Grubb's mounting resembles in principle that of Fraünhofer, though in the arrangement of its parts it is quite different.

The method pursued, and the formulæ used in adjusting the instrument, are given at length by Prof. B.; as also the mode of obtaining the value of a micrometer revolution.

The last part of the paper contains the results of the observations made on the comet of 1843, at West Point. Owing to unfavorable weather, the comet was observed only on the 24th, 25th, and 29th of March, and 2d of April.

The method pursued in making these observations was as follows: The difference of the instrumental places of two bodies, corrected for the change of instrumental errors arising from a change of position, being equal to the difference of their true places; if some well-known star be observed the same evening with the comet, the true place of the latter becomes known. To keep the instrument as nearly as possible in the same bearings during the observations on both bodies, α Ceti was selected, and used throughout. To correct these observations for the change of instrumental errors, the following formulæ of Kreil, Mem. Astr. Soc., Vol. IV. p. 495, were used; they were also used in making the adjustments of the equatorial.

$$S = \sigma + \Delta\sigma + \lambda \, \mathrm{Sin.} \, (\Phi - S) \, \tan. \, \delta + \mu \tan. \, \delta + \nu \sec. \, \delta$$
$$P = \pi + \Delta\pi + \lambda \cos. \, (\Phi - S)$$

In which S is the true, and σ the instrumental hour angle, corrected for refraction; P the true, and π the instrumental polar distance, corrected for refraction; Δ, σ and Δ, π the index errors of the hour and declination circles, respectively; λ the distance in arc between the pole of the heavens and that of the instrument; ϕ the hour angle of the instrumental pole, estimated from the meridian to the west; μ the difference between 90° and the angle which the declination axis makes with the polar; ν the difference between 90° and the angle which the line of collimation makes with the declination axis; and δ the declination of the body.

In this way the following places were obtained for the comet :—

		h	m	s					
March 25th,	Right Ascension,	3	35	20.14	Dec.—	7°	48′	46.6″	= δ
29th,	do.	3	55	05.8	” —	6	48	58.	= ”
April 2d,	do.	4	11	51.95	„ —	5	58	43.1	= ”

which being converted into longitudes and latitudes, and cleared from the effects of aberration by the usual formulæ for a fixed star, gave,—

March 25th,	Longitude,	49°	17′	52″	Latitude, —	26°	18′	11″
29th,	”	54	52	35	”	— 26	32	35
April 2d,	”	59	37	38	”	— 26	37	42

The portion of the aberration due to the proper motion of the comet, was applied to the time, according to the method of Mr.

Gauss. The correction in the place of the earth for the effect of parallax, was disregarded in consequence of its small value, the first curtate distance being greater than unity.

The method by which the following elements were computed was that of Dr. Olbers, as given by Dr. Bowditch in the appendix to his Commentary on the *Mécanique Celeste.*

Longitude of the ascending node,	357° 41′ 49″
Inclination,	36 41 48
Long. perihelion,	261 31 47
Perihelion distance,	0.053774
Perihelion passage, Greenwich mean time, Feb. 26th, motion retrograde,	6018
Distance from the earth on the evening of 29th of March,	107,002,000 miles.
Approximate diameter of the nebulous envelope,	36,800 ″

The Communication of Professor Bartlett was accompanied by plans, elevations, and sectional views of the building for the Library and Philosophical Apparatus, prepared by Lieutenants Richard Smith and Eaton, and by drawings of many of the instruments described.

Mr. Richard C. Taylor presented and read a memoir "On the Geology of the North-east Part of the Island of Cuba, and on the Character and Prospects of the Copper Region of Gibara."

The earlier part of Mr. Taylor's paper is given to a description of the various rocks, found in the northeastern parts of Cuba. These are principally magnesian, though there are also some of the calcareous class, which are highly interesting. They are all in a greater or less degree metamorphic, and occupy a highly inclined position.

Two principal and two subordinate parallel chains of white modified limestone, range across this part of the island in an E. N. E. direction. All these calcareous mountains are isolated, picturesque, and of singular form, as if thrust up from the midst of the serpentines, and greenstones, and diallage, which surround their bases.

An important geological feature of this country consists in a well-defined anticlinal axis, of which the most northern chain of limestone mountains is the center, and which pursues the E. N. E. direction alluded to. This axis tilts off the entire series of stratified rocks on either side: those on the north descending in that direction at an average angle of about forty-five degrees, and those on the south side of the axis dipping southward toward the interior, at an average angle of sixty-five degrees. Mr. Taylor remarked that the longitudinal extent of this axis was not ascertained with certainty; but that it was at least coextensive with the utmost limits of his observation, which embraced forty miles. He is inclined to the opinion, that the area which lies north of the anticlinal axis, and forms a belt eight or ten miles wide bordered by the sea, has undergone a greater metamorphic change than the area which stretches to the southward. In both instances it seems to be certain, that the igneous changes and the local disturbances of strata were greatest in the parts that are nearest to the center of the axis.

The author next describes the prevailing geological character of the savanna or mineral region of Gibara. He details some peculiarities of the limestone mountains; their remarkable outline as seen from the sea, their peculiar precipitous sides and columnar appearance when viewed at a shorter distance on the land.

A description follows of the mountain of La Silla, and the peculiarities attending its caves; and of the novel appearance there, and at various elevations on the mountain to within fifteen feet of its highest crest, of a compact calcareous formation crowded with shells, which on investigation prove to be those of the land. This new geological formation, at first so inexplicable, is traced to the enormous accumulation of land shells, in a dead state, in the caves and fissures and recesses of the mountain; their envelopment, mingled with other animal exuviæ and the dung of myriads of bats; and their subsequent consolidation by means of the filtration and crystallization of carbonate of lime: thus, in the progress of time, by this stalagmitical process, some of these caverns have been filled with solid matter.

Another yet more interesting fact presented itself. Among these partially fossilized land shells, were perceived other testacea that were unquestionably marine. The difficulty of accounting for this latter fact was removed by degrees, on witnessing the transporting agents then in operation. These are neither more nor less than the soldier crabs, [*Pagurus,*] which, at certain seasons, repair to the sea coast from the interior, to seek the dead littoral shells, and to appropriate them as habitations. Loaded with these they climb up the steepest mountains and traverse the densest woods. In such situations one continually perceives marine univalves, many miles in the interior, to which they have been carried by these persevering appropriators. As the caves form the retreat of a variety of animals, who in dying leave their bones upon the floors, these also become incorporated with the testacea of the land and the ocean; all contributing toward the construction of a new rock,—a geological puzzle which at some future day may furnish abundant matter for the disquisitions of the learned.

After describing the ancient and existing coral reefs that border the island, the author proceeds to the history of the discovery of copper lodes in the mineral region of Gibara; a brief account of the works that have been commenced there; and a few details and analyses of the ores. The paper terminates with notices of the deposits of chrome ore, and of the auriferous beds near Holguin.

Professor Booth read a Communication by James P. Espy, Esq., "On the Law of Cooling of Atmospheric Air for various suddenly diminished Pressures."

If clouds are formed from the cold of diminished pressure in up-moving columns of air, as Mr. Espy has contended in his "Philosophy of Storms"; and if it be true, as he assumes, that all the phenomena of storms are dependent on the latent caloric of vapor or steam which is condensed in the storms themselves; it becomes important to ascertain accurately what degree of cold is produced in air by the sudden diminution of pressure.

This Mr. E. has attempted to do by experiments with his Neph-elescope,—an instrument, consisting of an iron flask, such as is used for holding mercury, with a stop-cock attached, an exhausting pump, and a bent-tube barometer gauge. The results are arranged in tabular form at the end of his paper: the mode of attaining and applying these, he illustrates by reference to the first experiment of the table.

Having pumped out air from the flask, till the mercury in the inner leg of the gauge stood at 48 quarter inches higher than in the outer, and waited till the air within had acquired the temperature of that without, 66°, he opened the stop-cock so as to admit air freely, and closed it again at the moment of equilibrium of pressure. The condensation thus produced within the flask raised the temperature above that of the room; and the cock remaining closed, the mercury gradually rose as the air cooled within, till it rested at 6 quarters of an inch higher in the inner leg than in the outer.

Applying to this experiment the law, that air at the temperature $t°$ will expand to double the volume if heated $448° + t°$, Mr. Espy determines that all the air in the flask at the end of the experiment was heated 25°.7; and by reference to the proportion between the air which entered to produce the condensation and that which was condensed, he concludes, that the condensed air alone would have been heated 40°.7 nearly, but for its intermixture with the other. Remarking, however, that a much greater degree of heat is developed by condensing air into one-half than is absorbed by expanding it to double its volume, he ascertains by computation that air of the temperature of 66° is reduced 38°.6 by a sudden change from a pressure of 30 inches of mercury to that of 30 − 12, or 18 inches.

In the same manner, taking the mean of the experiments, Nos. 7, 12, 15, 17, and 19, he finds that with the air at 66°, a sudden change in the pressure from 29.92 inches of mercury to 13.10 inches, effects a reduction of temperature of 57°: and from experiments Nos. 8, 9, 10, 11, 16, 18, and 21, that air of the same temperature would be reduced 85° by a change of pressure from 29.91 to 9.91 inches.

Mr. Espy has observed, that when the air has been reduced but little in density, the effect on the temperature as ascertained by experiment is greater than that indicated by calculation; and that where the reduction of density has been greater, the result is otherwise. We may safely say, he remarks, that where the rarefaction is not pushed above two-thirds, the cooling effect is not greater than the proportion given above; a conclusion which has a most favorable bearing on the theory of storms presented in his work on that subject. The paper closes with a brief outline of his theory, and the following table:—

Experiments with the Nephelescope on Air under diminished Pressure.

Number.	Date. May 1893.	Barometer.	Thermometer	Difference of pressure before opening, in quarter inches.	Difference of pressure after opening, in quarter inches.
1	13	30.00	66°	48	6.
2	13	30.00	66°	51.5	5.5
3	13	30.00	66°	47	5.5
4	13	30.00	66°	76	1.25
5	13	29.90	68	53.5	6.
6	13	29.90	69	60.5	6.5
7	13	29.90	69	67.75	6.5
8	13	29.90.	69.5	72.5	7.
9	13	29.90	70	75	6.6
10	13	29.90	70	78.5	7.
11	13	29.90	70	84.5	7.25
12	14	30.00	67	67.5	6.5
13	14	30.00	68	70.4	6.5
14	14	30.00	70	26	3.6
15	14	29.90	70	65.5	6.5
16	14	29.90	70	81	7.65
17	14	29.90	70	66.3	6.4
18	16	29.99	76.5	88	7.3
19	16	29.99	77	69	6.5
20	16	29.97	78	71	6.75
21	16	29.97	78	80	7.
22	18	30.24	64	23.5	3.3
23	21	29.72	66	93	7.

From these experiments it appears also, says Mr. E., that the greatest rise of the mercury after condensation never amounts to eight quarters of an inch; and that the maximum rise occurs nearly when the density of the air within the nephelescope is reduced to about one-third: in experiment 23d, where the exhaustion was pushed the furthest, the rise after condensation was only seven quarters.

Professor Ducatel, of Baltimore, communicated a paper embracing a general view of the physical geography and geology of the State of Maryland, in connection principally with its agricultural condition and resources, being one chapter of the work entitled, "Physical History of the State of Maryland," now in progress of publication by himself and John H. Alexander, Esq., C. E.

In this communication, the varied aspect of the State of Maryland, presented by the magnificent Chesapeake Bay, with its numerous arms reaching to the inmost portions of the eastern division of the State, which comprises two-thirds of its territory; the table lands that cover the region of primary rocks and afford a great variety of good soils; the mountain ridges and intervening fertile valleys; the great susceptibilities of the soils in all parts of the State to improvement; and the abundant natural resources that everywhere present themselves, are dwelt upon in refutation of the allegation so generally made in popular works on geography, of the comparative insignificance of Maryland as a productive member of the great confederacy. The *eastern shore* is shown to consist of something more than arid sand-hills and pestilential marshes; and the *western shore* not to depend exclusively upon the rich valleys of Frederick and Hagerstown for its supplies. The region of country so amply provided with mineral wealth, in the way of coal and iron ores, is shown to possess also a very productive soil, and an amount of water-power capable of putting into activity a most extensive industry, and of sustaining a numerous population. The agricultural and mineral resources of the State are supposed by the authors to be equal, if not to surpass, those of any other portion of the United States' territory of the same extent. The communication likewise exhibits the most prominent features in the Sylva, Flora, and Fauna of the State.

A Communication, entitled, "Notice of the Meteorological Observations now making at the Military Posts of the U. S., by G. Mower, M.D., Surgeon U. S. A.," was read by Dr. Emerson; who prefaced it by some remarks on the present system of simultaneous observations, and on the state of Meteorological Science in the United States.

This paper, which was communicated on behalf of the Medical Department of the Army, presents a narrative of the operations of that department in collecting meteorological data, and solicits the cooperation of scientific men throughout the country.

Previous to the year 1818, we possess no records of meteorological observations taken in the United States on an extensive scale. During that year the surgeons at the military posts were directed to keep regular records of the weather, and transmit them quarterly to the Medical Bureau at Washington. The earliest registers, thus transmitted, and on file in the Surgeon General's Office, are dated January 1819.

The merit of introducing meteorological observations into the army is due to the late surgeon general, Dr. Joseph Lovell. The instruments at first were only, however, the thermometer and vane, to which the rain gauge was added in 1836. The results of these observations, from 1820 to 1830 inclusive, have been already published, and it is the purpose of the Department to complete the series to the present time.

As points of observations for the study of climate, our military posts possess peculiar advantages. They extend from the twenty-fifth to the forty-sixth degree of latitude, and from the sixty-seventh to the ninety-sixth degree of longitude; an immense region, embracing the whole inhabited area of the United States. A large proportion of them stand at intervals along two nearly parallel lines, running from south to north, and forming our eastern and western frontiers; extending on the Atlantic from Key West, near the Bahamas, to Passamaquoddy bay; and on the west, from the Balize to the Falls of St. Anthony. From this disposition of our posts, it happens that almost every parallel of latitude intersecting our country passes through a military station. And as the mean annual temperature of these posts has been determined by a series of observations, extending from ten to twenty years, we have been enabled to draw isothermal lines, coinciding at one or more points with almost every parallel of latitude, between Cape Sable and the St. Lawrence.

The United States, forming a zone of about twenty degrees of latitude, exhibit a range of mean annual temperature of 32°, which is equal to the range between Stockholm and Grand Cairo, a zone of 29°. The isothermal lines which intersect the capitals of Sweden and of Egypt, pass near Fort Brady, Mich., and St. Augustine, Fa., at a distance of only about 16°.

On passing the 30th degree of latitude, the climate of America is colder than that of Europe in the same parallels; and the difference increases in our progress from south to north. Thus Fort Monroe, Va., has the same mean temperature as Naples, 31° farther north; Washington, as Nantes, 8 1/2° farther north; Fort Wolcott, Newport, R. I., as London, 10° farther north; Fort Preble, Portland, Maine, as Edinburgh, 12° farther north; and Fort Sullivan, Eastport, Maine, as Stockholm, 15° farther north.

Since the beginning of the present year, the observations have been conducted on a larger scale, under instructions, matured by a board of which the author was a member, and subsequently approved by the present surgeon general, Dr. Lawson, and by the Secretary of War. Observations are now taken with the thermometer, the barometer and attached thermometer, and on the wind, the clouds, the clearness of the sky, and the dew-point by the wet bulb thermometer. These observations are recorded at four different periods of the day, except those with the wet bulb, which are taken only twice. The time when rain began and ended, and the quantity which fell, are noted at the close of every shower. In addition, the medical officers are requested to note under the head of "Remarks," all remarkable phenomena, especially sudden and simultaneous changes of wind and temperature; their effect on the barometer; the moment of greatest depression of the barometer in the passage of storms; the time of clouding; currents of clouds moving in different directions, and at different heights; the rise and fall of rivers and lakes; remarkable tides; the opening and closing of navigable waters; the last killing frost that occurs in spring, and the first in autumn, as shown by their effects on the tender buds, leaves, and germs of fruit trees, &c.; the commencement and progress of vegetation; the first appearance and departure of birds of passage; thunder storms, near or remote; silent lightning, with its direction and elevation above the horizon; falls of hail, snow, and sleet; fogs; white or hoar frost, &c.; also to examine the heavens at the latest hour of observation, whether there be any aurora, or shooting stars; and, especially about the 10th of August, and 12th and 13th of November, to see whether there be any great number of luminous meteors visible, stating the number observed in an hour, or at least in a quarter of an hour; and further, in cases of great fires occurring in clear, calm, dry weather, with a high dew-point, to observe whether clouds form over the fire, and to describe the phenomena.

At the equinoxes and solstices, *hourly* observations of the barometer are directed to be taken for twenty-four hours, to correspond with those already instituted at numerous points of Europe and America, at the suggestion of Sir John Herschel. The days fixed upon for these observations are the 21st of March, June, September, and December. But should any one of these 21st days fall on Sunday, then the observations will be deferred till the 22d. The observations at each station will commence at 6 o'clock A.M., and be continued at the beginning of each hour till 6 o'clock A.M. of the following days, care being taken to obtain the correct time.

The periods recommended for meteorological hours by the Royal Society of London, are 3 and 9 A.M., and 3 and 9 P.M. These periods the Department have adopted, with the substitution of sunrise for 3 A.M., as being more seasonable, and better suited to the routine of military service. The maxima and minima of the barometric oscillations, at the level of the sea, probably occur at these hours over a large portion of the globe. Besides, the lowest degree of temperature and dew-point are obtained, as experience has shown, shortly before the dawning of day, and the highest degree nearly at 3 P.M. The pair of hours, 9 A.M. and 9 P.M. coincide nearly, not only with the maxima of atmospheric pressure, but with the periods of mean morning and evening temperature; and half the sum of these two observations will be a near equivalent for the mean daily temperature, as is shown by Captain Mordecai of the corps of engineers, in his observations at Frankford Arsenal, near Philadelphia.

Observations are now made at sixty military stations, which with few exceptions are situated either on our maritime or inland frontier. The following list will give their distribution through the different states and territories: 5 in Maine, 1 in each of the states of New Hampshire, Massachusetts, Rhode Island, and Connecticut, 9 in New York, 4 in Pennsylvania, 2 in Maryland, 1 in Virginia, 2 in North Carolina, 1 in South Carolina, 2 in Georgia, 4 in Florida, 2 in Alabama, 5 in Louisiana, 3 in Arkansas, 1 in Arkansas Territory, 3 in Missouri, 1 in Missouri Territory, 4 in Iowa, 2 in Wisconsin, and 5 in Michigan.

But, the author remarks, while from the position of our posts the points of observation are thus extended along the frontier, the interior of our land is left comparatively destitute. On the Atlantic slope only two barometers are placed at any considerable distance from the ocean, at Carlisle Barracks, Pa., and Augusta Arsenal, Geo.: and there is not a single point of observation in the large tract of territory, embracing the states of Ohio, Indiana, Illinois, Kentucky, Tennessee, and Mississippi. The Department would therefore cordially invite the cooperation of colleges, scientific institutions, and individual admirers of meteorology in these sections of our country. The directions and forms used in the army, may at any time be obtained by applications addressed to the Surgeon General, U. S. A., Washington. All contributions from other sources will be cheerfully acknowledged in the publications of the Department. Under the direction of the Surgeon General, arrangements have been made for prosecuting meteorological inquiries with renewed diligence, and a medical officer will shortly be detailed to give his undivided time and attention to the subject, to arrange and digest the matter collected, and to prepare the results for publication.

Professor Booth read a communication by Mr. Henry C. Lea, of Philadelphia, entitled, "Description of some New Fossil Shells from the Tertiary of Virginia."

Mr. Lea observes, that the Atlantic tertiary of the United States from the St. Lawrence to Maryland is in patches, and from Maryland to the Gulf of Mexico in one broad sheet. It is considered as belonging near to the eocene, miocene, and post-pliocene periods of Mr. Lyell. Of these, the eocene has no recent species, while the post-pliocene has no extinct ones. The miocene is supposed to have about 17 per cent. of existing species, but Mr. Lea thinks that standard too high. For instance, at Petersburg, Va., there are found 68 species already described, besides 110 which he considers as new; in all 178. Of these, but nine are still existing; which gives us almost exactly five per cent. When we know more of our fossil conchology, the general per centage will most probably be reduced in the same manner. These shells frequently bear a remarkable similarity to those from the miocene of Dax, near Bordeaux, where the per centage of recent shells is between 30 and 40.

The shells described in the paper were obtained through the kindness of M. Tuomey, Esq., of Petersburg. Nearly all of the minuter species were detected by carefully examining a small portion of the marl from the vicinity of that place, as well as the sand scraped from the interior of larger shells, which is a favorite resort with some species. Among them Mr. Lea has found but two forms which appeared to require the erection of new genera.

The following list gives the names of the shells, with their classification, as indicated by Mr. Lea in his paper; the limits of this publication necessarily excluding the full descriptions by which he has characterized them.

FAMILY SERPULIDÆ.
GENUS SERPULA: S. convoluta, S. anguina.
GENUS PETALOCONCHUS (*Nobis*): Descr. Gen. *Testâ tubulari, solidâ irregulariter contortâ, laminis longitudinalibus duabus internis.* P. sculpturat us.

FAMILY TUBICOLIDÆ
GENUS TEREDO : T. catalpas, T. fistula.
GENUS GASTROCHANA : G. ligula.

FAMILY PHOLALIDÆ.
GENUS PHOLAS : P. rhomboidea.

FAMILY SOLENIDÆ
GENUS SOLEN : S. magnodentatus.
GENUS PANOPŒA : P. dubia.

FAMILY MYIDÆ

GENUS MYA : M. reflexa.
GENUS THRACIA (*Leach*): T. transversa.
GENUS ANATINA : A. tellinoides.

FAMILY MACTRIDÆ.

GENUS ALIGENA (Nobis) : Descr. Gen. *Testâ æquivalvi, subæquilaterali, posticè et anticè clauscâ; cardine dente cardinali uno, sulco sub natibus longo, minimè profundo.* A. striata, A. lævis.
GENUS KELLIA : K. triangula.

FAMILY LITHOPHAGIDÆ

GENUS SAXICAVA : S. oblonga.
GENUS PETRICOLA : P. compressa, P. bullata.

FAMILY NYMPHIDÆ

GENUS PSAMMOCOLA (*Blainville*): P. leporina, P. lucinoides, P. regia.
GENUS LUCINA: L. punctulata, L. lens.

FAMILY CONCHIDÆ.

GENUS ASTATE (*Sowerby*): A. lineolata.
GENUS CYTIIEREA : C. elevata, C. sphcerica.
GENUS VENUS : V. ascia.

FAMILY CARDIIDÆ

GENUS HIATELLA: H. lancea.

FAMILY ARCIDÆ.

GENUS NUCULA: N. dolabella, N. diaphana, N. æquilatera, N. carinata, N. acutidens.

FAMILY MYTILIDÆ.

GENUS MODIOLA : M. Spinigera.

FAMILY MALLEIDÆ.

GENUS AVICULA : A. multangulata.

FAMILY PECTENIDÆ.

GENUS PECTEN : P. micropleura, P. tenuis.
GENUS PLICATULA : P. rudis.

FAMILY PHYLLIDIDÆ.

GENUS CHITON : C. transenna.
GENUS PATELLA : P. acinaces.

FAMILY CALYPTRŒIDÆ.

GENUS CEMORIA (*Leach*): C. oblonga.
GENUS CALYPTRÆA : C. pileolus.
SUB-GENUS INFUNDIBULUM (*De Montfort*): I. concentricum.
SUB-GENUS CREPIDULA: C. ponderosa, C. cornucopiæ, C. lamina.

FAMILY BALLÆIDÆ.

GENUS BULLA : B. cylindrus.

FAMILY MELANIDÆ.

GENUS PASITHEA (*Lea*): P. exarata, P. subula, P. eburnea,
P. lævigata, P. ovulum, P. diaphana, P. turbinopsis, P. ornata.

FAMILY NERITIDÆ.

GENUS NATICA: N. aperta, N. spbmrulus, N. crassilabrum.

FAMILY PLICACIDÆ.

GENUS ACTÆON : A. granulatus, A. globosus, A. turbinatus, A. angula-
tus, A. glans, A. sculpturatus, A. nitens, A. milium, A. simplex.
GENUS PYRAMIDELLA: P. suturalis, P. elaborata.

FAMILY SCALARIDÆ.

GENUS SCALARTA: S. acicula, S. cornigera, S. micropleura. GENUS
DELPHINULA: D. costulata, D. concava, D. lipara, D. ob- lique striata,
D, trochiformis, D. glohulus, D. aperta, D. naticoides.

FAMILY TURBINIDÆ.

GENUS ROTELLA: R. sub-conica, R. carinata, R. lenticularis,
R. umbellicata.
GENUS TROCHUS: T. armillus, T. conus, T. lens, T. torquatus,
T. aratus, T. Ruffinii.
GENUS TURBO: T. glaber, T. rusticus.

FAMILY CANALIFERIDÆ.

GENUS CERITHIUM: C. clavulus, C. curturn, C. dmdaleum,

C. moniliferum.

GENUS PLEUROTOMA: P. lunatum.

GENUS FASCIOLARIA : F. parvula.

GENUS FUSUS: F. pygmmus, F. anomalus.

FAMILY PURPURIDÆ.

GENUS BUCCINUM: B. Tuomeyi, B. pusillum, B. frumentum, B. quadrulatum.

GENUS NASSA: N. impressa.

FAMILY COLUMELIDÆ.

GENUS MARGINELLA: M. cornulus, M. exilis.

GENUS OLIVA: O. canaliculata, O. ancillariæformis.

Professor Henry presented and read a communication "On a New Method of determining the Velocity of Projectiles."

The new method proposed by the author, consists in applying the instantaneous transmission of an electrical action, to determine the time of the passage of the ball between two screens, placed at a short distance from each other, on the path of the projectile. For this purpose, the observer is provided with a revolving cylinder, moved by clock work at the rate of at least ten turns in a second; and of which the convex surface is divided into a hundred equal parts; each part therefore indicating in the revolution the thousandth part of a second. Close to the surface of this cylinder, which revolves horizontally, are placed two galvanometers, one at each extremity of a diameter; the needles of these being furnished at one end with a pen for making a dot with printers' ink on the revolving surface.

To give motion to the needles at the proper moment, each galvanometer is made to form a part of the circuit of a galvanic current, which is completed by a long copper wire passing to one of the screens, and crossing it several times, so as to form a grating, through which the ball cannot pass without breaking the wire, and thus stopping the current. During the continuance of the galvanic action, the marking end of the needle is turned from the revolving cylinder, a few degrees, and pressed immovably against a "steady pin" by the well-known deflecting power of the electrical current; but the moment the current is stopped by the breaking of the long conductor, in the passage of the ball through the screen, the marking end of the needle is projected against the cylinder by the action of a fine spiral spring, similar to the hair spring of a watch, coiled around the center pin which supports the needle, and having an elastic force a little less than the deflecting power of the electrical current. The relative position of the dots thus formed gives the time of the passage of the ball through the space between the screens, and indicates the velocity at this part of the course.

The degree of deflection of the needle can be increased or diminished, by turning a screw, which alters the position of the "steady pin"; and the tension of the spiral spring can also be changed by an arrangement like that of the regulator of a watch.

In order that the position of the dots on the surface of the cylinder may exactly indicate the required interval of time, it is necessary that the time occupied by each needle, in starting from test and moving across the small arc to strike against the cylinder, should be precisely equal. If this be not the case, then the difference of these times will be the error of the instrument. This must, however, be exceedingly small, since the whole range of the end of the needle need not be more than the 20th of an inch; and the precise amount of error can readily be determined by experiment.

To adjust the apparatus for use, the galvanometers must be so placed that the two dots may be impressed on the cylinder, diametrically opposite each other when the instrument is at rest. The cylinder being then put in motion, the two circuits of long wire are placed together, so that they can be broken at the same instant by lifting a wire common to both from a cup of mercury. If, after breaking the circuits, the dots are still found in the same relative position, no further adjustment or correction will be required: but if this is not the case, then the springs may be altered until the dots are found in their proper positions; or the difference may be noted, and this constantly applied in each actual experiment as an index error.

To prevent the dot from the first galvanometer being confounded with that from the second, the two instruments are placed one below the other, in different horizontal planes.

In order that the pen may not describe a line on the cylinder, reentering into itself, and thus obliterate the dot first impressed; it may be found necessary to give the cylinder a slow ascending motion, so that a spiral instead of a circle would be marked on its surface. A chronometer for measuring minute portions of time, with a motion of this kind, is described in *Young's Natural Philosophy*, Vol. I. page 191.

To prevent agitations of the air, the whirling apparatus with the galvanometer may be placed in the vacuum of an air pump; and that part of the conducting wire, which crosses the screen, may be separated at each crossing, the ends being again united by slightly twisting them together, and the conduction being preserved by proper amalgamation, so that the force necessary to break the circuit may not sensibly lessen the velocity of the ball.

[Various other methods may be devised, for impressing a mark on the revolving cylinder, at the moment of the rupture of the galvanic current by the passage of the ball through the screen. But the following, which has suggested itself to Prof. H. since the meeting of the Society, and has been communicated by him to the Reporter, may be regarded as among the best. It dispenses with the galvanometers, and produces the mark by a direct electrical action.

A part of the long wire, which leads to the screen, is coiled around a bundle of soft iron wire; and over this is coiled another long wire; so as to produce an intense secondary current, on the principle of the common coil machine. One extremity of the secondary circuit is connected with the axis of the cylinder, and the other is made to terminate almost in contact with the revolving surface, which in this modification of the instrument is surrounded by a ruled or graduated paper. It is obvious, that the secondary current, which is induced by the interruption of the primary circuit, will pierce or mark the paper band at the moment of the screen being broken. There is no difficulty in effecting such a current of sufficient intensity to mark the paper: since Prof. H. in some of his experiments on Induction has developed one, which gave a spark between a point and a surface, of nearly a fourth of an inch in length.

The terminal points of the wires from the two screens may be placed very near each other in the same horizontal plane: if then the cylinder, revolving horizontally, has at the same time a slow ascending motion, the relative position of the dots on the paper will give the number of whole turns and parts of a turn, made by the cylinder while the ball was passing between the two screens. In the same way, the terminal points of wires from a number of different pairs of screens may be made to impress their marks on the surface of the same cylinder, and the velocity of the ball at the different points of its path may in this way be determined by a single experiment.

REPORTER.]

Mr. Kane read a letter, addressed to him by George Bancroft, Esq., dated Boston, 22d May, 1843, expressing his regrets at being absent from the Society's meeting, and communicating the following extracts from unpublished letters of Dr. Franklin:—

Benjamin Franklin to Cadwallader Colden.
Philadelphia, Feb. 13th, 1749–50.

Sir,—I received your very kind letter, relating to my proposals for the education of our youth, and return you the thanks of the gentlemen concerned, for the useful hints you have favored us with. It was long doubtful, whether the academy would be fixed in the town or country; but a majority of those, from whose generous subscriptions we expected to be able to carry the scheme into execution, being strongly for the town, it was at last fixed to be there. And we have, for the purpose, made an advantageous purchase of the building which was erected for itinerant preaching: a house one hundred feet long, and seventy wide, with a large lot of ground, capable of additional buildings, situate in an airy part of the town. It cost, I suppose, not less than £2000 building: but we have it for less than half the money. It is strongly built of brick; and we are now about dividing it into rooms for the academy. The subscription goes on with great success, and will not, I believe, be much short of £5000, besides what we expect from the proprietors. From our government we expect nothing. Enclosed I send you a copy of our present constitutions; but we are to have a charter, and then such of the constitutions as are found good by experience will, I suppose, be enacted into laws, and others amended, &c.

In this affair, as well as in other public affairs I have been engaged in, the laboring oar has lain and does lie very much upon me.

* * * * * * *

I have no observations of Jupiter's satellites to send you, as I expected I should have. Being myself otherwise engaged, and not very skillful in those matters, I depended on our astronomer, Mr. Godfrey, and put the telescope into his hands for that purpose. He had a fine summer for it, but * * * * * * * * * * * * our surveyor general Mr. Scull, who was his neighbor, could never get him to assist in making the meridian line. He is now dead, and your letter of directions for making such a line, which I put into his hands, is lost. Mr. Scull desires me to write to you for a repetition of those directions, and when you have a little leisure I shall be obliged to you for them; but it will now be midsummer, before we shall have an opportunity of observing Jupiter again.

I wrote some additional papers on electricity, which I will get copied and send to you per next post. They go on much slower in those discoveries at home, than might be expected.

I am glad you are about enlarging and explaining your principles of natural philosophy. I believe the work will be well received by the learned world.

Benjamin Franklin to Cadwallader Colden.
Philadelphia, Feb. 28th, 1753.

* * * * * We are preparing here to make accurate observations on the approaching transit of Mercury over the sun. You will oblige us much by sending the accounts you have received from Lord Macclesfield of his great mural quadrant. I congratulate you on your discovery of a new motion in the earth's axis. You will, I see, render your name immortal.

I believe I have not before told you, that I have provided a subscription here of £1500, to fit out a vessel in search of a north-west passage: she sails in a few days, and is called the *Argus*, commanded by Mr. Swaine, who was in the last expedition in the Calefornia [*sic*], author of a journal of that voyage, in two volumes. We think the attempt laudable, whatever may be the success. If he fails, *magnis tamen excidit ausis.*

Mr. Thomas Gilpin laid before the meeting some fine specimens of the Bombax of Santa Cruz.

SPECIAL MEETING.

Eighth Session, 30th May, half past 5 o'clock, P. M.

Dr. PATTERSON, Vice-President, in the Chair.

Mr. Ellwood Morris made an oral communication relative to the TURBINE of Fourneyron, a horizontal hydraulic motor first employed in France in 1827, and of late successfully introduced into use in our country by Merrick & Towne of Philadelphia.

Mr. Morris traced the history of this machine from the first suggestion of Belidor, supported by M. Navier. He adverted to the prize of 6000 francs, offered in 1 827, 1829, and 1832, by the Society for encouraging national industry, to the person who should successfully project and put in use two hydraulic motors, which should not waste water, but receive it into their buckets without shock and cause it to leave them without velocity. He alluded to the experimental researches of M. Poncetel and of M. Bardin; the latter of whom employed wheels with vertical axes, and gave them the name of *Turbines*, but without fully complying with the prize conditions. These were attained by M. Fourneyron in 1833, after ten years devotion to the subject, by the construction of two turbines; one of which realized a useful effect of more than 80 per cent of the power expended, and the other with less expenditure of water than an overshot wheel, which it supplanted as the motive of a blowing machine at a furnace, furnished a greater blast.

Mr. Morris exhibited a full-size horizontal section of a turbine, just constructed by Merrick & Towne for the powder-works of Dupont in Delaware, and a smaller well-constructed model, which exposed all the parts of the machine. He explained its theory and mode of action, and made copious references to Fourneyron's *Memoir in the Bulletin of the Society* for encouraging National Industry for 1834, and to the *Comptes-rendus* of the Academy of Sciences, the *Journal of the Franklin Institute,* and the works of Morin, for information both as to its theory and the economy of its results.

Mr. Morris next adverted to an apparent error in the conclusions drawn by Fourneyron, from the theory of the turbine, in consequence of his having considered merely a single filament of water, in its progress along a curved guide, and through the wheel, instead of applying his reasoning to the line of motion of the center of gravity of the issuing particles. M. Fourneyron had thus been led to prescribe a rule for proportioning the tangent angle of the curved guides, which generally gave them at the inner circle of the wheel an inclination of about 45°, with a radius through the same point, instead of an angle of 70°, and upwards, which results from applying the theory to the acting line of the central filament of the moving mass.

In order that the water, agreeably to one branch of the theory, shall act upon the wheel without shock, it is only necessary to cause the final element of the curved bucket recipient of the hydraulic force, to be the resultant of two forces; the one representing the direction and velocity of the center of gravity of the water issuing from the directing sluices, whilst the other is a tangent to the inner circle of the wheel, and indicates its motion at the point of entrance of the acting fluid.

To cause the water, conformably to the other branch of the theory, to leave the wheel without velocity, the bucket must be curved around sufficiently to receive the full pressure that the head of water is capable of imparting, yet not so much as to obstruct the egress of the fluid when it has done its work; to which end it appears to be necessary, that a tangent to the final element of the curved bucket shall intersect a tangent to the outer circle of the wheel, with an angle opening outwards, of from 10° to 15°.

The curve necessary to satisfy these conditions, has been carefully determined by actual experiment; and Mr. M. remarked, that an advantageous form of the machine having been once devised, it serves as a *type* for constructing others, adapted to a different fall and volume of water. Thus, in the language of M. Combes, a French Engineer of Mines, to whom this generalization is due,—"knowing the fall and the volume of water to be expended, for a wheel to be constructed, we make it similar to the type; its linear dimensions will be those of the type, directly as the square roots of the volumes of water expended, and inversely as the fourth roots of the heights of fall; its angular velocity will be to that of the type, directly as the fourth roots of the cubes of the heights of fall, and inversely as the square roots of the volumes of water expended."

The experiments made in France with the brake of M. Prony have established, that the coefficient of effect of turbines, or the ratio of power actually realized to that expended is at an average seventy per cent. Mr. Morris has recently tested this result at the Rockland Mills in Delaware, where the turbine is employed to drive a cotton mill: his experiments are collated in the table, which closes this abstract.

From these tabulated experiments it will appear, that with lifts of sluice gate ranging from 5 to 7 inches, or from 5/8 to 7/8 of the full height of the turbine, and with velocities at the inner circle, varying from about 4/10 to near 6/10 of the theoretical velocity, due to the working fall of water, this motor realized a useful effect, varying from 64 to 70 per cent of the absolute power expended, or of that which is theoretically due to the expenditure of water and the available fall at the time.

The maximum effect seems to have been derived when the lift of the sluice gate equaled 6 inches, or two-thirds of the full height of the wheel; and when the turbine at its inner circle ran at a speed equivalent to 46 per cent of the theoretical velocity of the water, issuing under a head equal to the working fall.

An examination of the experiments from the 6th to the 14th inclusive, will show that the coefficients of effect within these limits, notwithstanding considerable variations in the relative velocities or the wheel, and its impelling water, averaged 67 per cent: thus showing that this turbine when run, with a strong lift of sluice gate, realized as high a coefficient of effect as was assigned by Smeaton to overshot water wheels.

With regard to the following table, Mr. M. remarked that the quantity of water used, which fixes the theoretical power due to the expenditure and descent, was determined by applying to the openings of the directing sluices certain coefficients of discharge, deduced from those of Morin on the turbine of Mullbach by a comparison of the velocities and lifts of gate in the one and the other case: the results therefore are merely proximate, but cannot be very distant from the truth.

The total fall of water at the Rockland Mills is usually about seven feet; but the turbine has continued to drive the machinery of the mill effectively when the difference of level was reduced by backwater to three feet three inches, and the wheel was entirely submerged to the depth of four feet. With an external diameter of $4\,^2/_3$ feet and a vertical thickness of about 8 inches, it propels the same machinery that heretofore required two breast wheels, one of 14 feet bucket and 10 feet diameter, the other of 8 feet bucket and 16 feet diameter, and uses one-third less water than the latter of these alone.

Mr. Morris next adverted to the durability of the turbine: he supposes it less liable to wear at the pivot than the common water-wheel; as the latter while running supports a heavy load of water, from which the turbine is relieved by the interior fixed disc, which carries the directing sluices. In the turbine at Rockland, the pivot is ingeniously lubricated with oil by a syphon wick passing through an opening in the center of the vertical shaft; after five months' use, the wear of the pivot is not perceptible.

Mr. M. concluded by exhibiting and describing the original model of an *inverted* turbine, devised by Mr. Young of the Rockland Mills in 1840; and added that a machine, similar in all essential respects, had been contrived by Mr. Erskine Hazard in 1842, without a knowledge of Mr. Young's. Mr. M. mentioned some objections to which he thought the inverted turbines would be liable; and observed, that for the reason he had given while speaking of the durability of the ordinary turbine, he believed that the pressure, and consequently the friction would be the same in all the proposed varieties of the machine and modes of running it.

Dr. Meigs made some remarks upon Cyanosis Neonatorum, and upon a new and more successful mode of cure.

Experiments upon the Turbine at the Rockland Mills near Wilmington, Del., made by Ellwood Morris, Civil Engineer, January 21st, 1843.

1	2	3	4	5	6	7	8	9	10	11	12	13	14	15
No. of the Experiments.	Vertical lift of the annular sluice gate in inches.	Aggregate average of the fixed disk openings exposed by the sluice gate in sup. feet.	Proximate coefficients of discharge of the fixed disk openings.	Depression of the water surface in the head and tail races below a fixed level. In Head race.	In Tail race.	Working fall during the experiments.	Theoretical velocity of water due to the working fall in feet per second.	Proximate quantity of water discharged by the Turbine in cubic feet per minute.	Theoretical power in horses' power of 33,000 lbs. lifted one foot high per minute.	No. of turns of the wheel per minute.	Load of brake at the end of a radius of 11 feet.	Horses' power developed of 33,000 lbs. lifted 1 foot high per min.	Ratio of the speed of the wheel at the inner circle to the theoretical velocity of the water due to the working fall.	Approximate coefficient of effect, or ratio of the power actually realized, to the theoretical power.
1	1	0.344	.930	3.083	9.937	6.854	20.9	401	5.2	38	14	1.1	.317	.211
2	2	0.687	.923	3.083	9.625	6.542	20.5	779	9.6	58	28	3.4	.500	.354
3	3	1.031	.918	3.125	9.479	6.354	20.2	1146	13.8	72	42	6.3	.621	.456
4	4	1.375	.830	3.188	9.271	6.083	19.8	1355	15.2	74	56	8.6	.652	.566
5	5	1.375	.724	3.125	9.187	6.062	19.7	1471	17	59	70	8.6	.505	.688
6	5	1.719	.742	3.125	9.187	6.062	19.7	1507	17.3	57	98	11.7	.593	.682
7	6	1.719	.720	3.220	9.208	5.979	19.6	1745	19.8	67	84	11.8	.565	.662
8	6	2.062	.720	3.229	9.208	5.979	19.6	1745	19.8	64	98	13.1	.525	.672
9	6	2.062	.700	3.188	9.125	5.937	19.5	1688	19	59	108	13.3	.411	.690
10	6	2.062	.700	3.188	9.125	5.937	19.5	1688	19	46	136	13.1	.465	.700
11	7	2.062	.700	3.125	9.042	5.917	19.4	1960	22	52	122	13.3	.445	.641
12	7	2.406	.700	3.104	8.958	5.917	19.3	1950	21.6	49½	136	14.1	.510	.667
13	7	2.406	.700	3.146	8.958	5.854	19.3	1950	21.1	56½	122	14.4	.542	.683
14	7	2.406	.700			5.812				60	115	14.4		

To illustrate these, he exhibited a magnified model of the fœtal heart; in which were shown the auricular cavities, with their septum; the foramen ovale covered, on the left side, with its valve; the Eustachian valve, and the passages to the ventricles, with the great vessels, &c.

Dr. Meigs remarked, that the representation showed the Eustachian valve, springing from the anterior column of the arch of the foramen ovale, and extending to the anterior limb of the circular orifice of the inferior cava, and demonstrated the truth of Winslow's rationale of the fœtal circulation, viz:—that the blood of the inferior cava passes mainly across the auricle, and raising the valve upon the left side of the septum, enters the left auricle, passing thence to the left ventricle and the systemic distribution, without reaching the pulmonary branches. The model showed further, that the blood of the upper cava falls into the auricle opposite to the iter ad ventriculum, passes most readily through that opening, and that there is therefore a crossing of the currents.

It is not rare to meet with new born children, especially with those that are premature, in whom this crossing of the currents continues after birth. Any considerable degree of this decussation, Dr. Meigs remarked, involves of necessity the production of the state called *cyanosis,*—a state in which the entire mass of blood becomes loaded carbon. Asphyxia, more or less complete, is the consequence of the failure to eliminate the carbon and to absorb the oxygen. This asphyxia is Cyanosis, or Morbus Cœruleus.

Dr. Meigs had seen a great many patients die under these circumstances. The books contained no rationale for a philosophical treatment, and he was much at a loss to discover a successful one. He found upon reflection, that the valve of the foramen ovale is lifted by the current from the inferior cava, projected against it by the sides of the Eustachian valve, and also by the gravitation of the blood in the right auricle if the child be lying upon its left side. These reflections he made, while in presence of an infant apparently in the agonies of death from cyanosis. He laid it on its right side, with its head and shoulders inclined upwards on pillows, and requested that it should remain for several hours in that position. The success of this mechanical treatment was perfect.

Upon placing the child thus, its septum auricularum became a horizontal plane, supporting the blood in the left auricle. The weight of that blood pressed the valve of the foramen ovale into coincidence with the plane of the septum, and closed the patulous orifice. The succeeding injections of blood took their proper route to the lungs and the system; so that a few acts of the respiratory muscles gave sufficient doses of oxygen to the blood to diminish and rapidly to remove the excess of carbon; and the child was cured. This treatment, Dr. Meigs stated, had been successful under his administration of it, in rescuing from impending death upwards of twenty persons. It begins, he said, to be understood and practiced extensively in Philadelphia, and other parts of the United States. But he was desirous to take advantage of the present convention of the Philosophical Society, at which so many members of the profession were present, to exhibit his model, and to make these explanations of a mechanical treatment of a before unmanageable and fatal disorder.

Prof. Bache gave an account of the observations made at Philadelphia and Toronto, during the magnetic disturbance of May 6, 1843, and pointed out their bearing upon the question of the kind of instruments and observations appropriate to determine the phenomena during rapid changes of the magnetic elements.

The disturbance was first noticed at Philadelphia between 3 and 4 P.M. (9 and 1,0 P.M. Göttingen time), when observations at term-day intervals were commenced. At 12 hours, Göttingen time, observations were begun upon the declinometer at every two minutes (the mean time corresponding to the even minutes), and upon the horizontal force magnetometer at alternate intervals of two and four minutes, and generally at 0, 4, 6, 10, &c. minutes after each hour. The vertical force magnetometer was observed every six minutes, viz. at 2, 8, 14, 20, &c. minutes after the hour, throughout the disturbance. The instruments were most disturbed between 9 $^1/_2$ P.M. of the 5th, and 1 A.M. of the 6th of May, Göttingen time, and again between 4 $^1/_2$ and 8 A.M. Göttingen time. From about 2 A.M. to 6 A.M., and again from about 7 until 8 A.M. Göttingen time, observations were made upon a small horizontal force magnetometer, and upon a Lloyd inclinometer placed in a building not far from the Observatory. The extremes of vibration in each were noted, so that the mean time of the several observations of the horizontal force instrument corresponded to 0, 2, 6, 8, 12, &c. minutes from the time of beginning, and of the vertical force to 4, 10, 16, &c. minutes.

A similar disturbance was perceived at Toronto, Canada; where the instruments were observed at the term-day intervals from about 10 P.M. Göttingen time, of the 5th of May, until midnight at Toronto. The declination magnetometer was thus recorded at 0, 5, 10, 15, 20, &c. minutes after each hour, the horizontal force magnetometer at 2, 7, 12, 17, &c. minutes, and the vertical force at 3, 8, 13, 18, & c. minutes after each hour. Prof. Bache owed to the kindness of Lieutenant Younghusband, R. A., director of the observatory at Toronto, and to the liberal character of the instructions for the government of the British observatories, the communication of the observations made at Toronto, and the permission to use them.

Prof. Bache had hoped that similar results might have been obtained from the magnetic observatories at Cambridge and Washington, but no special observations had been made there. He regretted to learn from Professor Peirce that the observations at Cambridge had, in fact, been discontinued.

The instruments in the Observatory at the Girard College are of the largest dimensions in use, the declinometer and the bifilar being of Gauss's pattern, and the vertical force magnetometer upon a corresponding scale. The time of vibration of the instruments, respectively, is 24, 45, and 30 seconds. The Toronto instruments are of the comparatively moderate dimensions of Professor Lloyd's pattern. The subsidiary horizontal force instrument at the Girard College is of still smaller dimensions, being $9 \, ^3/_4$ inches long; its time of vibration is about ten seconds. It was to the comparative results obtained with these instruments of different dimensions and very different times of vibration, that Prof. Bache wished particularly to direct attention. In a letter received from Col. Sabine, dated May 1st, 1843 (extracts from which Prof. Bache read), a résumé is given of the opinions of the leading magneticians of Europe upon this question, showing that contributions toward its determination are required. This letter contains, further, a discussion of the appropriate intervals of observation, in reference to which the comparative results now presented showed that the intervals at present used are probably incompetent to give an accurate representation of the phenomena.

The following comparisons were illustrated by referring to the broken lines, traced in the usual manner for representing the changes in the different instruments.

Declination.—During a small portion of the first part of the disturbance, when the Philadelphia observations were made at intervals of six minutes, and those at Toronto at intervals of five, it was difficult to judge how far the apparent discrepancies in the movements of the magnets were real, though some of them were probably so. When the observations at Philadelphia at intervals of two minutes began, it was at once apparent that the rapid movements were dissimilar. Commencing the comparison at 24 hours (12 P.M.) Gottingen time, the needle appeared to move, at Philadelphia, steadily eastward from 24 hours to 24 hours 6 minutes; while at Toronto, at 24 hours 05 minutes it appeared to be to the west of its position at 24 hours. At 0 hours (May 6, 24 hours May 5) 16 minutes, a very small minimum occurred at Philadelphia, between two maxima at 0 hours 14 minutes and 0 hours 18 minutes; while at Toronto, at 0 hours 15 minutes, there was a very low minimum. At 0 hours 25 minutes there was a maximum, and at 0 hours 30 minutes a minimum, at Toronto, neither of which appeared at Philadelphia. During this part of the disturbance the changes of declination appear to have been very rapid, and the apparent coincidences of phenomena at the two places are not satisfactory. The intervals between the observations were obviously too great to represent the phenomena accurately. Were the phenomena really dissimilar? or will the difference in the dimensions of the instruments explain the discrepancies?

During the second great movement in this disturbance, which began about 3 hours 50 minutes, A.M., of May 6th, Göttingen time, the motion of the needle was less rapid than during the first, and the results at Philadelphia and Toronto agree better, though between 5 hours 40 minutes and 5 hours 58 minutes, they are quite discrepant.

Excluding this period, there were seven apparent maxima out of fourteen, the times of which agreed precisely with those at Toronto; one only of these not being very clearly defined. Five others were within two minutes, the differences of time being fairly referable to differences in the epochs of observation. One was without a corresponding maximum at Toronto, the discrepancy arising probably from the same cause; only one was a real discrepancy. The correspondence in the times of occurrence of minima of declination was as close as that of the maxima. The extent of the corresponding movements in the two places was, however, by no means proportionate in different cases. It would appear from these results, that when the changes in the declination were not rapid, the large bar gave the same epochs and directions of change as the smaller one, which renders it probable that the discrepancies observed in more rapid changes were due to the difference of instruments rather than to real differences of phenomena. This, however, Professor Bache remarked, is a point which further observation would more satisfactorily determine.

Horizontal force. The curves representing the period of the first considerable change of horizontal force at Philadelphia and Toronto, presented striking discrepancies, with one remarkable coincidence, that of the greatest movement of increase. The maximum was reached at 23 hours 44 minutes, Göttingen time, at Philadelphia, and 23 hours 47 minutes, at Toronto, these being the nearest corresponding periods of observation. Between 23 hours 22 minutes and 23 hours 44 minutes, there was an apparent increase of horizontal force at Philadelphia, of .017 (428 divisions of the scale of the instrument), and at Toronto of .027 (306 divisions of the scale).

It was during the second period of disturbance that the small bifilar instrument was observed at Philadelphia as well as the large one. Taking the periods of maxima and minima as shown by the broken lines it appeared, that when the results given by the Gauss instrument at Philadelphia were compared with those of the Toronto instrument, in twenty-six cases, ten agreed, six were doubtful, and ten disagreed. When the corresponding epochs, as shown by the small horizontal force instrument at Philadelphia and the Toronto magnetometer, were compared in seventeen cases, thirteen agreed, two were doubtful, and two disagreed. The range of movement of the small bifilar was but about one-sixth that of the large one, and in general the means of accuracy of observing inferior; and yet the coincidence of its results with an instrument approaching to its dimensions is very striking. The strongest feature in this part of the disturbance was a decrease of force between 5 hours Göttingen time, and 5 hours 55 minutes. The Gauss bifilar at Philadelphia gave the amount of change .015 of the horizontal force, the small bifilar gave .034 and the Toronto instrument .025. In three corresponding changes of smaller amount, the three instruments taken in the order just named gave, respectively, for the first change .0039, .0099, .0085; for the second change, .0012, .0025, .0057; for the third change, .0016, .0036, .0036.

Vertical force.—The results of a comparison of the movements of the vertical force instruments at Toronto and Philadelphia were shown to confirm those deduced from the comparison of the bifilars, though it was remarked that the inference is less unexceptionable than in the former case, because the amount of friction at the axes of the two instruments is very different.

Professor Bache, stated, in conclusion, that it was his intention to pursue this subject, by mounting a set of small magnetometers to be observed during disturbances, and by observing the Gauss bifilar and perhaps the declinometer, at each vibration, on the same occasions, whenever it might be practicable.

Dr. Goddard called the attention of the Society to the experiments of Moser, which had been supposed to prove the existence of invisible photographic rays.

He said, that in repeating these experiments with much care, he had entirely failed to obtain an image. This failure he ascribed to the fact, that before attempting the supposed photographic process, he had made the cameo or coin which was to produce the image, and the plate on which it was to be received, *perfectly clean.* On mentioning the circumstance to Mr. Joseph Saxton, of the U.S. Mint, whose expertness in experimenting is well known to the members of the Society, he learnt that numerous and careful trials had proved to his satisfaction, that the effect remarked by Moser was due to the evaporation of some greasy substance from the surface of the object forming the image, and that when this had been first carefully removed, no image was obtained. This had recalled to Dr. G.'s mind an observation which he made some years ago, while prosecuting a series of experiments on the Daguerreotype. He had wrapped some highly polished plates in a very old newspaper for the night, and found in the morning that the outer plates had received a perfectly distinct image of the printing that had been in contact with them. Thinking that this was owing to the contact, he enclosed some similar plates in very fine and clean tissue paper, and wrapped the newspaper over this; but the impression appeared in the morning as before, the oil having traversed the tissue paper. This led him to the precaution of employing tin boxes to keep the plates when made ready for the Daguerreotype process.

Dr. G. concluded by expressing his entire concurrence in the opinion of Mr. Saxton, that the effects observed by Moser were due to the evaporation of oily or other organic substances which had accumulated on the surface of the body forming the image, and that the agency of heat was important only as it facilitated the evaporation. He added, as his belief, that all substances evaporated at all times and under all temperatures; the only difference being in the rate: in one case it was inappreciable from its slowness, in another distinguished readily by the phenomena which it produced.

In the course of his remarks, he alluded to the first employment of bromine in the photographic process, and exhibited the first Daguerreotype specimen produced by means of it. It was made in Philadelphia, by himself and Mr. Cornelius, in December 1839.

The remarks of Dr. Goddard led to a free conversation on the subject, in which Mr. Saxton, Professor Henry, Professor James Rogers, and other gentlemen, took part. In the course of it, the following account was given of Mr. Saxton's experiments.

A gold coin, half an eagle, which had been dipped in pure nitric acid, then washed in distilled water, and afterwards dried by whirling in the air, was placed on a well-prepared Daguerreotype plate, and suffered to remain undisturbed for four days. At the end of this time no impression was visible when the plate was breathed on, except at two spots corresponding to the opposite sides of the coin where it had been grasped by the wooden pincers when plunged into the acid.

A copper coin was next placed above a Daguerreotype plate, with nothing between them but an exceedingly thin plate of mica, which had been split from the middle of a thick piece. But after so remaining for three days, no impression of the coin could be observed, though the mica was found, by actual measurement, to be less than the one-thousandth of an inch in thickness. The same coin, placed on the same plate, without the interposed mica, gave an impression in the course of four hours; and when the coin was slightly warmed, a like effect was produced in one hour.

To determine if there was any difference in the screening effect of different substances, a thin plate of sulfate of lime was next placed between the coin and the prepared plate, and the whole suffered to remain five days: at the end of this time, however, no image could be perceived.

In another experiment, a thin plate of glass was interposed between the coin and the plate, with the same negative result. The experiments were also varied by using different metals; but in no case were any effects produced through the thinnest transparent substance which could be procured.

That this was not due to the distance of the coin from the plate, was evident from the fact, that when the former was supported, by pieces of mica under its edges, at the same distance as in the last experiment, an image of the part of the coin not screened by the mica was impressed on the plate, while no such effect was produced by the parts under which the mica was placed.

As, then, those parts of the coin which are either perfectly clean or which have been thus screened give no image, the conclusion is, that the effects observed by Moser are due to the evaporation of the volatile matter which has infilmed the coin. Some parts of the coin, such as the salient points of the figures, would be differently soiled from the others, and would also evaporate the volatile or fatty matter differently; and when the coin is placed very near a polished surface, the condensation of the evaporated matter on this surface would be different at different points, and present the appearance of an image.

The principle of the formation of these images may be simply illustrated by slightly touching the point of the finger to a clean plate of glass. If the plate be afterwards breathed on, the vapor will be differently condensed on the parts which have been in contact with the raised lines of the skin; and hence an image of the surface of the finger will be exhibited on the glass. If the finger could be held at the distance of the one-fiftieth of an inch for a few hours, the same effect would be produced by the unequal vaporization.

The same conclusion has also been arrived at by M. Fizeau, and was communicated to M. Arago in November last; but the investigations of Mr. Saxton were entirely independent of any knowledge of the French experiments, and his explanation of the phenomenon had been communicated to Professor Henry and other members of the Society, before any account of the experiments of M. Fizeau reached this country.

Professor Rogers also mentioned, that he had repeated some of the experiments of Mr. Saxton at the time, and had been fully convinced that his explanation of the images was the true one.

Prof. H. D. Rogers made an oral communication, under the title of "Geological Notices."

He alluded first to the subject of Coprolites, which had been brought before the Society by Mr. Lea, and stated that he considered the specimens, which he had discovered in the green sand formation of New Jersey, to belong to the fossil Crocodile, with whose bones he found them. He did not know whether those previously collected by Dr. De Kay appertained to the same animal; but thought it probable they did.

He then passed to the subject of the paper read by Mr. Taylor, the flora of the carboniferous period. The generalization first fully announced by Logan, that each seam of coal reposes on a peculiar indurated mud, called fire-clay, in which abounds a characteristic plant, the *Stigmaria ficoides*, has been of late extensively confirmed by observations in this country, made by Logan, Lyell, and some of our home geologists. Prof. R. read a letter from Mr. Weld to Prof. Ducatel, which goes to show, that in one or two cases of apparent exception to the rule, in the Potomac coal-field, a critical reëxamination of the strata has resulted in the detection of the fire-clay and its plants, in close contact with the under side of the coal. Prof. Rogers thinks that each bed of coal is an ancient stigmaria bog; slow decomposition by moisture, and exclusion of air, having caused the conversion to coal.

He then proceeded to apply the doctrine of earthquakes, contained in the paper by his brother and himself, to the explanation of certain geological phenomena in the Appalachian chain of the United States. He showed by reference to a general section and several pictorial views, that each anticlinal flexure of this mountain region exhibits, with few exceptions, a more rapid arching on its N. W. side, or the side remotest from the region of greatest subterranean disturbance; and that as we recede to the N. W., the plication of the strata lessens, the curves flatten out, and the dips on the opposite sides of the axes approach more to equality, until a general horizontality succeeds in the plain of the Ohio and its tributaries. A similar configuration, and law of progressively diminishing incurvation in the flexures, prevails it is believed in all regions of the earth's surface, where a distinctly developed system of anticlinal axes can be studied. The axes, or great flexures, of the Appalachian chain, occur in natural groups; those in each group being approximately parallel, and equidistant, and coinciding nearly in length. Some of these are more than 100 miles long. The distance separating two adjacent axes in an extensive group, visibly augments as we cross them toward the N. W., in which direction the flexures, it has been said, flatten out. Professor Rogers and his brother think they have proved, that analogous relationships connect together the axes of the more disturbed regions of Europe, and that a few simple laws will be found sufficient to express the phenomena of flexures in every part of the globe.

Connecting these general facts respecting the form, parallelism, and progressively increasing distance, and flattening of the flexures, with the manifestations of analogous transient bendings of the earth's crust during earthquakes, they have suggested that all the phenomena may be united under one theory of dynamic movement. They conceive, that in those districts of the globe where the strata are much undulated, the crust was subjected to an excessive upward tension, or other internal force, causing in it extensive linear disruptions and all the phenomena of earthquakes on the grandest scale. Simultaneously with its violent oscillation, the crust would be shoved horizontally forward in the direction of the transmission of the earthquake waves, and the undulations or flexures be rendered permanent by being braced or keyed fast by the intrusion of the molten lava below into the innumerable fissures which would open. The horizontal or tangential movement would steepen the advanced side of each anticlinal wave, and every fresh pulsation would augment the amount of flexure, so that even a complete plication with folding under would finally result on that side of the region, whence these repeated and violent earthquake pulsations proceeded.

Prof. Rogers concluded by citing in support of his declaration, that the structural phenomena of the Appalachian chain extend to Europe, a letter that he had recently received from a distinguished British geologist, Prof. Phillips.

Professor Stephen Alexander, of Princeton, N. J., presented a communication "On the Physical Phenomena which accompany Solar Eclipses."

In this communication, Professor Alexander brings forward various considerations in support of the conclusions announced by him to the Society on the 15th of July last;* and as prefatory to the argument which he founds on them, he gives a synopsis of numerous observations of the phenomena in question made subsequently to the invention of the telescope. These he arranges in tabular form in the order of dates, under titles following the natural order of

---

* See Proceedings A. P. S. Vol. II., May, June and July, 1842, p. 201, or the summary at the end of this abstract.

succession of the phenomena. The observations thus detailed and collated, are necessarily reserved for the forthcoming volume of the Transactions of the Am. Phil. Society: the following is little more than a succinct catalogue of the titles under which they are classed by Prof. Alexander.—They form his FIRST TABLE.

1. An appearance of a doubtful character, similar to that of a faintly illuminated limb of the moon, seen by Prof. A. in the State of Georgia, shortly before the beginning of the eclipse of November 30, 1834, and not unlike that which was observed by J. L. Memes, Esq. in England, to succeed the eclipse of September 7, 1820; both observations apparently indicating that a portion of the moon's disc was visible.

2. The appearance of a prominent dark point or points, indenting the sun's disc at the very beginning of the eclipse; observed by Sir Wm. Herschell in 1793, M. Mechain in 1794, and by others in 1836 and 1838;—and corresponding phenomena at its ending, observed by Herr Schmidt and Herr Gutkaes at Dresden in 1818, by Mr. Memes in 1820, by Mr. Rümker in 1828, and by others in 1836; and at a later period of the eclipse, an exaggerated roughness of the moon's edge, observed particularly by Prof. Weidler in 1730 and 1733, and by others in 1765, 1793, and 1820, &c. &c.

3. An appearance of dark lines on the sun's limb soon after the first disturbance; observed by Prof. Johnson at Philadelphia, in 1838;—and corresponding phenomena near the end; observed by Professor Van Swinden, in 1820, and by Prof. Johnson, Dr. Patterson, Mr. Justice, and Mr. Walker, in 1838.

4. A slight bending of the cusps of the uneclipsed portion of the sun; noticed by the Prince de Croy in 1765, Dr. Herschell in 1793, and Mr. Greve in 1820.

5. A peculiar motion or agitation of the sun's light, at the edge of the moon's disc; referred to by Mr. Maclaurin as having been seen in 1736–7, and previously observed by Hevelius; observed also by Prof. Weidler in 1739 and at other times.

6. The appearance of flashes, or coruscations of light, upon the moon's disc; observed in 1715, by Dr. Halley at London, seen also at Wittemburg in 1739, and by Mr. Memes in 1820.

7. An illuminated band or seemingly brighter portion of the sun's disc, bordering the limb of the moon; observed by the author,

February 12, 1831, at Berlin, in the State of Maryland, during the progress of the eclipse, with two telescopes of different powers and differently colored screen-glasses.

8. A remarkable illumination of a solar spot "close to the moon's edge"; observed by Mr. Dawes of Ormskirk, and by Messrs. Lasell and King of Liverpool, May 1836.

9. An isolated spot or trace of light upon the moon's disc:—the spot, observed about the middle of the eclipse of July 14, 1748, by Mr. Short at Aberdour Castle,—and during the eclipse of September 7, 1820, for nearly $1/2$ min. by Mr. Memes:—the trace or stream of light also, seen by Mr. Memes, extending from each side of the macula; its entire length about 60° of the moon's circumference.—A faint but somewhat similar light was observed also by the author on the moon's disc shortly before the total obscuration, Nov. 30, 1834.

10. Projection of the visible edge of the moon's disc beyond the solar cusps; observed at London, in June 1666, by Mr. Willoughby and others; at- Nuremberg in Sept. 1699, and at different places in 1715, 1748, 1816, 1820, and 1831.

11. Peculiar color of large continuous portions of the moon's disc: —its edge pale like clouds, the middle of the disc black, as described by Wolf in 1706: black with a blue girdle, as seen by Prof. Weidler in 1727: the same appearance with an interior girdle of reddish purple, seen by the same observer in the eclipse of 1730: the blackness of the disk tinged with red, as seen by him July 1739: a dusky grey orb, somewhat brighter toward the sun, seen by M. Schroeter in 1793: reddish purple, gradually darkening to the center, which was black, as seen by F. Baily Esq. and Mr. Veitch during the existence of the annulus, in May 1836, with refractors and red screen-glasses.

12. Illumination, with greater or less distinctness, of a crescent of the moon's disc: observed by M. de Ferrer at Kinderhook in June 1806, just before the emersion from the total eclipse; the color palish yellow: observed also by Mr. Garnett at New Brunswick, N. J. Mr. Hassler at Washington in Feb. 1831, noticed the "inequalities of the moon" by the "reflected light and shade" upon the $1/3$ of its disc when the moon was about immersed toward the beginning and end of the eclipse.

13. Greater or less illumination of the entire disc of the moon at the time of total eclipse: it appeared uneven and rough, and of

a saffron hue to Prof. Waller at Upsal in 1715: Vassenius at Gottenburg, in 1733 saw the spots in the disc: M. Arago at Perpignan, and Mr. Airy on the Superga in 1842, saw the whole disc illuminated.

14. The appearance of separate portions of the sun's disc, usually in the form of drops of mercury, with intervening spaces of greater or less extent; these spaces sometimes seeming to be afterwards drawn out into narrow dark lines. These phenomena having been seen only when the uneclipsed portion of the sun began to present very fine cusps: noticed by Prof. Heinrich at Breslau in 1706, by Halley at London in 1715, by several observers in Scotland in 1736–7, at Elgin in 1748, and at different stations in 1764, 1780, 1791, 1806, 1820, 1831, 1834, 1836, 1838, and 1842.

15. Red color of a small fragment of the sun's disc, the only portion remaining uneclipsed: remarked by Mr. Shelton near Pontefract, April 22, 1715.

16. An apparent distortion of the moon's disc on the same side with the dark lines already described (No. 14), and cotemporaneous with them: observed by F. Baily, Esq., 15th May, 1836.

17. The sudden formation of the annulus, which at once had sensible breadth: remarked whenever the *drops* had been seen, and the eclipse was not far from central.

18. The comparatively faint light of the remaining fragment of the sun's disc just before it suffered a total eclipse; noticed by Dr. Halley and others at London, April 22, 1715, and Messrs. Pinnaud and Boisgeraud at Toulouse, July 8, 1842,—and the seeming disappearance of such a portion of the disc, and its subsequent reappearance just previous to its total immersion; noticed by Mr. Airy upon the Superga, July 8, 1842.

19. A succession of shadows upon the surface of the landscape, "like the rippling waves upon some shallow pond," seen about the time of the total eclipse of June 16th, 1806, by Mr. Crookshank and others in the interior of the State of New York: also the progress of the shadow; which was observed by M. Duillier at Geneva in May 1706; according to Dr. Haley, at Brightling, Sussex Co., in April 1715; by M. Lorenz, in Lemberg, Nov. 1816; at Beaufort, S. C. by R. T. Paine, Esq., Nov. 1834; at Turin in 1842 by Professors Plana and Forbes.

20. A corona of variable extent, and essentially, it would seem, of a white color, surrounding the moon during a total eclipse, and

in some rare cases when the eclipse was not quite total; when most extensive, the outer parts of the corona less brilliant than the inner: observed at several places on the continent of Europe in May 1706, at London and Upsala in April 1715, at Gottenburg in 1733, and also at several stations in 1778, 1806, 1816, 1834, and 1842.

21. Various colors exhibited by the corona:—ruddy bright, and as a ring of gold, at Zurich, in 1706, to Scheuzer and to Wurzelbaur at Nuremburg; pale white or pearl, a little tinged with colors of iris, to Dr. Halley in 1715; reddish, changing to pale yellow; pearl color; and nearly white, with a tinge of peach color; to different observers in 1778, 1806, and 1842.

22. Apparent motion in the corona: a fluctuation or trepidation about the limb of the moon at Barcelona in 1628, seen also by Tschirnhausen at Dresden in 1706; a whirling motion, seen by Don Ulloa in 1778; a clear unsteady light, according to Von Stölpe, in 1816; vivid and flickering, as seen by Mr. Baily in 1842.

23. The appearance of small isolated spots or beams of light, colored or otherwise, at various points of the moon's edge, at the time of a total eclipse, often of exceeding richness and beauty of tint; variously described as "reddish spots," "red flames," "illuminated Alps," " fiery mountains," "rocks of incandescent crystal"; remarked by Vassenius at Gottenburg, May 3, 1733; and with very interesting variety by several observers in America and Europe, Nov. 30, 1834, and July 8, 1842. Like phenomena were seen between the cusps, the eclipse being in most cases nearly annular, by Mr. Short at Aberdour Castle, July 14, 1748; and in Sept. 1838, by the author and Prof. Henry at Princeton, N. J., Mr. Gummere at Haverford, Pa., Prof. Johnson, Mr. Walker, and others at Philadelphia, with red, yellow, and greenish-yellow screen-glasses. In this last eclipse, Dr. R. M. Patterson at Philadelphia saw a beam of light between the cusps for four minutes fifty-three seconds after the rupture of the ring: the same was noticed at Princeton by the author and Prof. Henry, through a Dollond telescope with red screen glass, but was not seen when a much superior instrument of Utschneider and Fraunhofer was used, of which the screen-glass was greenish-yellow.

24. "Very slender columns of smoke" apparently issuing from the western limb of the moon during a total eclipse, as observed by M. de Ferrer at Kinderhook, N. Y., June 16, 1806 (perhaps a modification of the foregoing, No. 23).

25. The appearance of a brilliant spot within the moon's disc after the total immersion of the sun, and previous to its visible emersion at the moon's limb; noticed by several observers in 1778, 1806, and 1842.

26. Several luminous appearances, each of which has at some time preceded the sun's emersion from a total eclipse,—and the analogous phenomena which have been seen when the eclipse was all but annular:—such as the increase of light in the atmosphere for about two seconds before the sun's limb was visible, noticed by Dr. Bowditch and others in June 1806,—the ribbon of white light concentric with the sun, nearly fifteen seconds before the emersion observed by M. de Ferrer in June 1806, and about eleven seconds before the emersion observed by the author in Nov. 1834,—a narrow band of red light occupying the same position,—a bright gleam of light, sometimes red, sometimes purple, between the cusps, when the eclipse was nearly annular, sometimes before the formation, and sometimes after the dissolution of the annulus, all of which have been often observed.

27. The peculiar color of the sky, the appearance of the clouds, &c., and other indications of the loss of light during the progress of a total eclipse, and of the character of that light which remained; particularly described by Dr. Halley in the instance of the eclipse of April 22, 1715; by Messrs. Pinnaud and Boisgeraud at Toulouse, July 8, 1842; by the author in Nov. 1834, &c. &c.

28. A manifest increase of light soon after the total obscuration of the sun on June 24, 1778, and a subsequent and apparently corresponding decrease previous to the sun's emersion at the edge of the moon's disc; witnessed by Don Ulloa.

29. Indistinctness of the prominent points at the edge of the moon's disc near the end of an eclipse, and an apparent adhesion of the disc to the surface of the sun,—followed by a sudden termination of the eclipse,—described by J. L. Memes, Esq., in his account of the eclipse of Sept. 7, 1820. The two latter phenomena were also seen by Mr. Andrew Livingstone at Gibraltar.

30. Change in the relative extent and brilliancy of the different colors of the solar spectrum,—the yellow and blue rays being in one instance increased in brilliancy, and the red becoming faint and of reduced breadth; observed at the eclipse of Sept. 7, 1820,

at Norwich, Eng., and by Mr. Paine in Massachusetts at that of Feb. 12, 1831: Messrs. J. and S. R. Gummere at Burlington, N. J. thought that the violet, though very bright, was rather paler than at other times.

31. Effects attributed to irradiation and inflexion;—observed repeatedly by different astronomers.

32. Heat of the sun's rays which remained unobstructed; and various effects upon the temperature, pressure, &c., of the atmosphere, as observed during sundry eclipses of notable size;—including very numerous experiments with burning glasses, observations with Leslie's photometer, indications of the thermometer exposed to the sun and in the shade, and of the barometer, observations on the deposition of dew and the variations of the dewpoint, the condensation of vapor, and other atmospheric changes.

33. Magnetic variations; observations of the needle by Col. Beaufoy, Sept. 7, 1820, Mr. Gummere at Burlington, N. J., Feb. 12, 1831, Mr. Nicollet at Milledgeville, Geo., Nov. 30, 1834.

Professor Alexander exhibited drawings on a large scale of several of the appearances described under these heads, and presented a list of several publications in which others are represented. He remarked that some of the observations which he had noticed, might be regarded as of little value, and that some were probably referable to ocular deception; that perhaps too, separate places had been sometimes assigned to phenomena which were only modified effects of the same cause; while others had been grouped, because presenting themselves at like periods and seemingly resembling each other, which might yet prove to be essentially different. He observed that he had kept one uniform rule in view: viz. *not to reject any thing which might be of use; but when it rested on sufficient evidence, to present it distinctly, free, as far as possible, from the bias of any preconceived notion, with regard to its reality or its cause.*

He observed that in accounting for many of the phenomena, it should not always be his object to advance a satisfactory theory, or perhaps even a stable hypothesis, but to attain to what might be termed an incipient generalization, which should so group the phenomena, that though he might fail to point out their cause, he might at least show wherein such cause must reside.

He adverted in the first place to the supposed action of a lunar atmosphere as the most obvious cause of some of the phenomena, and inquired how far such an atmosphere would be favorably or unfavorably situated for producing them. As the moon always presents nearly the same face to the earth, whose mass far exceeds her own, an aëri-form fluid covering the entire surface of that satellite would be subjected to a great terrestrial and anti-terrestrial tide, whose position would be subject to but little variation. The tide producing energy at the moon would be about 158 times as great in its intensity as the like action of the moon upon the earth. The action of the sun would also combine with that of the earth at new and full moon; so that the result of both would be to accumulate a great tide wave above the central portion of the disc, and another on the opposite invisible side; thus leaving the edges comparatively destitute, and most so at the time of an eclipse. If therefore the moon be surrounded by an atmosphere, we are most unfavorably placed for the study of its effects.

The question of the existence or non-existence of a lunar atmosphere, however, is not a primary one in this connection; and Professor A. proceeds to inquire, what are the laws which regulate the action of the moon upon the light passing near her border. To aid in determining these, he forms a SECOND TABLE of the peculiar phenomena which have been observed to accompany certain occultations of planets and fixed stars. These he classes under titles and with an arrangement similar to those observed in the table of phenomena of solar eclipses. The following is a condensed abstract of these titles:

1. An apparent projection of a fixed star, or of a planet or some portion of a planet on the moon's disc; remarked at two occultations of Venus by La Feuillée in 1722, and M. Thulis at Marseilles in 1796, at the occultation of Jupiter and his satellites by Mr. Ramage at Aberdeen in April 1824, at an occultation of Uranus by Captain Ross in August 1824, at twenty-six occultations of fixed stars recorded in Sir J. South's table, in Vol. III. Part II. of the Memoirs of the Astronomical Society of London, and at occultations of Aldebaran and other fixed stars, observed by the author and others.

2. A distortion of the disc of a planet, or of the false disc of a star: observed at the occultation of Venus in Sept. 1729, by C. Kirch at Berlin; of Jupiter in April 1824, by Mr. Ramage at Aberdeen, and by the author and others at different times; of Saturn in Oct. 1825 by R. Cornfield and J. Wallis; and of fixed stars in several instances.

3. A division of the illuminated disc of a planet into two parts, and their subsequent reunion at the instant of complete emersion; witnessed by M. Thulis at the occultation of Venus, Nov. 1799.

4. Apparent adhesion of stars to the moon's edge; of which twenty cases are recorded in Sir J. South's table; observed too by Maraldi in 1718, Cassini, Da la Hire, and others.

5. The gliding of a star along the moon's edge previous to its immersion: three instances of which are recorded by Sir J. South; observed too by M. Ideler, at Berlin, in 1792.

6. A tremulous motion of a star just before the immersion, as seen by the author for two seconds in the case of γ Tauri in 1831, or immediately after the emersion as seen by M. Flaugergues, Sept. 1796.

7. A seeming emersion of some portion of the disc of a planet, and its subsequent occultation; observed in Russia by MM. de la Croyére and De l'Isle at the occultation of Mars, Jan. 1726.

8. An apparently sudden and violent motion in the star as it disappeared at the immersion, and a like rapid separation from the moon's limb at the emersion, observed in several of the cases already mentioned, in which the star was projected on the disc, and in one case by M. Ideler at Berlin in 1792.

9. An apparent diminution in the brightness of a planet or fixed star a little before the immersion and for a short time after the emersion. This was noticed by Capt. Ross in August 1824, at the immersian of Uranus, at the emersion of Saturn in June 1762, by Mr. S. Dunn, and at the immersion or emersion of fixed stars, five instances of which are mentioned in Sir J. South's tables, and others were observed by the author, Prof. Henry, and other astronomers.

10. Disappearance of a star at a sensible distance from the moon's limb previous to any ordinary immersion, observed by Dr. Maskelyne, Dec. 1779, and the corresponding phenomenon at the reappearance of a planet or fixed star; observed by Capt. Ross at the occultation of Uranus in Aug. 1824.

11. A change of color in the light of a star at the immersion or emersion; of which seven instances are noticed by Sir J. South and others.

12. A faint impression or brush of light just after a star had vanished at the immersion; observed by Dr. Maskelyne and by the author.

To this catalogue of phenomena attending occultations, Prof. A. adds a THIRD TABLE, embracing certain luminous appearances observed on various occasions upon the moon's disc. These may be abstracted thus:

1. A faint line extending beyond the sharp cusps of the illuiminated part of the moon's disc, at certain periods near the time of new moon; observed by Mr. Schroeter at Lilienthal, and ascribed by him to a lunar twilight.

2. A luminous arch completely uniting the cusps, soon after new moon; observed by Kolben in 1705 and by Siegesbee.

3. Luminous spots having a fiery appearance, which have at different times been seen upon the dark hemisphere of the moon; mentioned by Dr. Herschell and by M. Schroeter, and referred by them to volcanic action.

4. A peculiar band of light extending across a portion of a solar spot, which seemed to be inaccessible to the sun's rays; first observed by Bianchini in 1725, extending from the illuminated margin across the middle of the spot Plato, and subsequently in 1751 by Mr. Short, Dr. Stephens, and Mr. Harris, who saw two such streaks across the same spot; one of which was afterwards again divided: they also saw two gaps in the mountainous margin, directly interposed between the streaks and the sun.

5. Unusual tints of light, and tinted shadows, seen at certain spots, which at the time were situated near the terminator or boundary of lunar day and night; mentioned in Beer and Mädler's Selenographie, p. 153.

From these Prof. A. passed to phenomena attending the transits of Venus and Mercury, some of which have been recognized as bearing a strong resemblance to those observed during eclipses of the sun: These also he has arranged in tables. The following is a summary of his FOURTH TABLE, embracing phenomena noticed at transits of Venus:

1. "A kind of lucid wave" of an imperfectly transparent character, which gave the first intimation of the approach of the planet; observed by Mr. Dunn at Greenwich in 1769.

2. An apparent ferment or boiling following the foregoing phenomenon, at the same place of the sun's limb, and continuing for some seconds till the limb of the planet was seen to enter; also observed by Mr. Dunn in 1769, and with slight variations by Bevis at Kew, Hirst at Greenwich, and others.

3. Nearly cotemporaneous with this last appearance, a "watery pointed shadow" was seen by Dr. Smith at Norriton, a shadow with irregular notched edges by Mr. Shippen at Philadelphia, a dusky appearance by Dr. Williamson, and an obscuration gradually advancing by Mr. Ewing, both at the same place, a penumbra by Mr. Horsfall at London, &c.

4. The first visible impression of the dark body of the planet in most cases presented nothing remarkable.

5. A border of light around that portion of the planet which had not yet entered upon the sun's disc, was seen by Mr. Rittenhouse and other observers at Norriton and Philadelphia in 1769, by Dr. Maskelyne and others at the same transit, and in 1761 by M. Bergman and M. Mallet at Upsala.

6. A distortion of the portion of the planet's disc which was on the sun after partial immersion:—the notch appearing to Mr. Hirst at Greenwich, in 1769, to be rough, and as if a portion of a much less sphere; and the planet assuming an oval or elongated appearance, to Dr. Wilson at Glasgow, M. Ferner at Stockholm, and M. Mallet at Upsala.

7. A distortion of the cusps of the sun previous to the internal contact; observed by M. Mallet at Upsala in 1761.

8. "A partial and very faint illumination, both a little without and a little within the sun's limb, as well as in the limb itself, where the contact was to be," when the outer edge of Venus had come near to the sun, and the border of light around it had vanished;

observed in 1769 by Mr. Dunn at Greenwich and Mr. Canton at London.

9. A renewed ebullition, succeeded by the formation of a ligament, penumbra, or a black drop, &c., extending from the body of the planet, so as to conceal the sun's limb; thus delaying as it were the internal contact; noticed by Messrs. Hirst and Dunn at Greenwich in 1769, and with more or less accordance by very numerous observers in Europe and America: it was noticed also in 1761 at Upsala, Madras, and elsewhere.

10. A feeble illumination, with more or less violent agitation of various parts of the ligament above mentioned, which in one case seemed to consist of "many black cones or fringes"; particularly noticed by Mr. Dunn in 1769, and M. Mallet at Upsala in 1761; noticed also by others.

11. A continued agitation, brighter light, and the frittering away or dissolution of the ligament; described by Mr. Dunn in 1769, Dr. Smith at Norriton, and several others.

12. The completion of the thread of light: in 1769, as seen by Mr. Dunn, it succeeded immediately the changes last described; at Norriton, according to Dr. Smith, the points of the luminous thread closed, but at first with imperfect light; at Philadelphia the continuous thread had to Mr. Pearson a tremulous motion; Dr. Maskelyne saw a fine stream of light flowing gently round; Mr. Horsfall at London a lambent light whirling round the opaque limb; and the thread, when complete, appeared to different observers in 1761 and 1769 to be one-sixth, one-eighth, and one-tenth of the diameter of Venus in breadth.

13. A distortion of the disc of the planet after the dissolution of the ligament, when the whole disc was visible on the sun, was observed by Dr. Maskelyne at London in 1769, and at several other astronomical stations.

14. A glimmering light about the part of the planet which had last entered upon the six's disc; seen by Mr. Hirst, 1769.

15. A purplish-colored light on the northern side of Venus, which endured for six or seven minutes, was observed by Mr. Horsley at Greenwich in 1769.

16. A narrow luminous circle, seen at different times during the transit, while Venus was wholly on the sun; by Mr. Hitchins at

Greenwich and by Mr. Dunn in 1761, and M. Ferner near Paris and M. Lulofs at Leyden; the last of whom says, that it disappeared when the sun's light became intense enough to require colored glasses.

17. An eccentric ring of light around Venus, having a gentle fluctuating motion, seen by Mr. Mason at Cavan, 1769.

18. A seeming illumination of the disc of the planet, noticed by Mr. Dunn, 1761.

19. The reappearande of the black drop, &c. (noticed above, 9), at the second internal contact, was observed by Mr. Wales in 1769 for twenty-four seconds as a protuberance, as a penumbra by the Earl of Macclesfield, as a ligament at Upsala in 1761, and a black drop at Wardhus in the same year.

20. A protuberance of the edge of the sun's disc at the first external contact, surrounding Venus on all sides by its margin, which appeared to be a little elevated beyond the planet: followed moreover by a distortion of the cusps when the intervening thread of light had disappeared. This was observed in 1761 by M. Mallet at Upsala.

21. The distortion of a portion of the planet's disc, which still indented the sun after the first external contact; observed by MM. Stroemer and Mallet at Upsala in 1761.

22. Luminous appearance around a part of the planet, which after the first external contact was observed to extend beyond the sun's disc; noticed in 1761 by MM. Mallet and Bergman at Upsala, and by M. Wargentin at Stockholm.

23. A sharp indentation just before Venus suddenly left the sun; observed by M. Mallet at Upsala in 1761.

24. A somewhat faint penumbra or ligament, after the egress of the dark disc of the planet; observed by M. Bergman at Upsala in 1761.

Phenomena somewhat similar to those mentioned in the foregoing table have also been occasionally observed during the transits of Mercury; but they have been of comparatively rare occurrence. They have not escaped notice through inattention; as they were looked for very carefully in various instances, especially at the transit of 1769, which occurred but a few months after the last transit of Venus, and that of 1802, which was observed with scrupulous care and under most favorable circumstances by Dr. Maskelyne. The phenomena which have been exhibited at the transits of the smaller planet form Prof. Alexander's FIFTH TABLE.

1. The appearance of something like a penumbra at the first external contact; remarked by Prof. Williams at Harvard University, Nov. 12, 1782. The appearance was not that of the contact of two circles, nor like a well-defined black spot entering upon the sun, but rather like a dark oval shadow entering and mixing with the sun's limb.

2. The appearance of a black drop or ligament, or of a cohesion, previous to the completion of the thread of light at the first internal contact. (See Bode's Jahrbuch for 1803, pp. 109 and 198, and Mem. Astr. Soc. Lond. Vol. VI. p. 198.)

3. Apparently elliptical form of Mercury's disc when near the sun's limb,—noticed at or soon after the ingress, in Nov. 1723, and Nov. 1736, by several European observers, by Prof. Williams at Harvard in Nov. 1782, and by Dr. Madison of William and Mary College in Nov. 1789,—at or near the egress, at Nuremberg in Oct. 1690, at Vienna in Nov. 1697, at Louvain in May 1786,—at both the ingress and the egress, by President Willard of Harvard in Nov. 1782,—and at a period of the observation not recorded, in May 1799 by M. Flaugergues at Viviers.

4. A luminous ring around the disc of the planet; seen by M. Plantade in Nov. 1736 during the entire period of the transit and for some seconds thereafter, by Mr. Short in 1736 and 1743 as a luminous corona, by M. Prosperin in May 1786, by the assistant of Dr. Maskelyne at the same transit, by M. Eimbecke at Hamburg in May 1799, and by the author himself at Albany, N. Y. in May 1832, without a dark glass, through thin, though dark, flying clouds.

5. A dark nebulous ring around the planet; possibly a modification of No. 4; seen by M. Schroeter in 1799, by M. Ljunberg in 1802, and by Prof. Moll in 1832.

6. Peculiar appearances on the disc of the planet; described and figured by Wurzelbaur of Nuremberg, Nov. 1697, as a minute globule, not perfectly round and of graded intensity of light, darker at the lower section, very bright at the upper, and with shaded brightness between Prof. Mole saw an imperfectly defined spot upon the disc in 1832.

7. Reappearance at the egress of the black drop, or of an apparent adhesion; spoken by Prof. Jungnitz of Breslau and M. Kohler of Dresden.

8. A seemingly rapid motion, and sundry changes in the appearances, at the second internal and external contacts, as seen through various screen-glasses: Reference is here had particularly to the observations of M. de Barros made at Paris in May 1753, with an excellent Gregorian reflector, his smoked glass being fixed in a close tube, perpendicular to the axis of the telescope. After noting the interior contact through a green glass held over the smoked glass, he found, on using the smoked glass only, that a thread of light was visible between the limbs for four seconds after; the exterior contact too seemed stationary for six or seven seconds; and after the total egress had been seen through the combined glasses, Mercury was again seen on the sun through the smoked glass alone for six or seven seconds, &c. &c. Whether all these observations could have been completed in so brief an interval, has been seriously questioned.

9. A change of form in the indentation as Mercury was retiring from the sun's disc, followed by an adhesion, or by something like a fine dark ligament, which suddenly disappeared at the last external contact. In Nov. 1690, M. Warzelbaur observed this adhesion after the planet had recovered its roundness, and in 1697 this distortion and its sudden disappearance.

With reference to the question, whether the phenomena, described in the two tables immediately preceding, ought to be referred to the peculiar action of the planets themselves, Prof. A. turns to the phenomena which they have exhibited under other circumstances. Those of Venus, derived principally from the writings of M. Schroeter of Lilienthal, and Dr. Herschell, but in part from the works of others, and from his personal observations, form a SIXTH TABLE.

1. A manifestly superior brightness of the limb of the planet, which usually was observed to diminish with no little rapidity toward the interior edge of the illuminated surface,—a common observation of M. Schroeter; mentioned particularly by Dr. Herschell in 1793; seen by the author in Oct. 1839.

2. Appearance of a dark spot or spots upon the bright portion of the disc; observed by Fontana in 1645, Cassini in 1666, and others in 1726, 1727, 1728, 1783, 1799, 1820, and 1840.

3. Uneven edge of the terminator or boundary of day and night; observed by Fontana in 1645, Cassini in 1667 and 1668, and others in 1700 and 1822, and by the author in Oct. 1839.

4. Variety in the apparent form of the cusps of the enlightened portion of the disc; one appearing to be sharp and the other round, or both of them round; noticed especially when the planet was near to its greatest elongation;—observed at different times by M. Schroeter, M. Tischbein, and Dr. Chladni.

5. A bright projecting point in the terminator, near the northern round cusp; seen on one occasion by M. Tischbein and M. Schroeter.

6. Intrusion of a distinct dark point from the edge of the terminator into the illuminated hemisphere: also faint shades between the asperities of the terminator;—observed by M. Schroeter, Dr. Chladni, and Dr. Olbers, on different occasions. On one of these occasions, 27th Feb. 1793, the dark point, such as had been noticed near the southern cusp the day before, was seen by M. Schroeter to extend itself, "during the observation," entirely across the cusp; so that in eleven minutes the end of the cusp "passed very evidently to the form of" No. 7.

7. "A separate point of light" beyond the apparent end of the cusp:—seen on other occasions also by M. Schroeter.

8. Peculiar form of one or both of the sharp cusps of the enlightened portion of the disc, when the planet was near the inferior conjunction:—bent apparently inward in the form of a hook, as seen repeatedly by M. Schroeter in 1790, and by the author in Oct. 1839: —or turned outward in the direction of a tangent, as seen by the author in the same month.

9. Faint termination of one or both of the sharp cusps, which were not unfrequently prolonged beyond the limits of a hemisphere; observed and measured by M. Schroeter in 1789, 1790, and 1793, also by Dr. Herschell, and in 1839 by the author.

10. Apparent deviation from a regular curvature in other portions of the limb, at a time when Venus was near to the inferior conjunction; observed by M. Schroeter in 1790, and by the author and others in 1839.

11. Projection of a long continuous portion of the terminator beyond its general outline (a seeming exaggeration of No. 3),—noticed by M. Schroeter in 1793, and by Herr Pastorff in 1820.

12. Visibility of the whole edge of the obscure part of the disc, when the brightly illuminated part presented the usual appearance of a crescent; the planet being not very far from the inferior conjunction:—observed by Herr Pastorff.

Mercury also has occasionally exhibited phenomena of similar character; these Prof. A. arranges in a SEVENTH TABLE.

1. Two manifest inequalities in the outline of the limb near the southern cusp; resembling the mountain Dörfel, when it is seen on the moon's edge:—observed by M. Schroeter in 1800.

2. Shadows of mountains:—so regarded by M. Schroeter.

3. Roundness of the southern cusp; and contemporaneously —

4. A bending or other irregular form of the northern cusp; observed by MM. Schroeter and Harding in 1800.

5. Apparent variation in the intensity of the planet's light;—invisible on one day at the time of its meridian passage, though the weather was favorable; yet visible distinctly the day before and the day after; as remarked by M. Vidal in Jan. 1799;—whose observation is confirmed generally by M. Bode.

Having arranged in this manner the phenomena which are to be considered, Prof. A. passes to an inquiry into the origin of the broken cusps, dark spaces and lines, brilliant drops, &c., described in the 14th paragraph of Table I.

He adverts first to a suggestion of Mr. Airy, the English Astronomer Royal, in the account of his observations on the total eclipse of July 8, 1842. Mr. Airy had been looking carefully with the darkened glass upon the eye-piece, and had satisfied himself that the sun had entirely disappeared. He was proceeding to note the time, when his companion asserting that to him it was still visible, he recurred to the telescope, and again saw the narrow ring of the sun's disc, though not quite so bright as before, and saw the moon's limb again advance and cover it. The explanation which he proposes, but not very confidently, is, that the light of the sun's disc, very near to its limb, being considerably less than in those parts of the disc which are farther from the limb; the interference of a cloud, sufficiently dense to hide the faintest part of the limb but not to hide the brighter parts, would sensibly lessen the sun's diameter; and that in fact such a faint cloud was seen upon the sun at the time of his first apparent extinction, which, though not sufficient to conceal the edge of the sun's disc from the naked eye, might yet have concealed it from an observer through the telescope, in which the specific brightness is always less than it is to the naked eye, and which in the observed case was armed by a dark glass. Mr. A. adds, that before this he had referred the notches in the sun's limb to irregularity of refraction; but that these, and perhaps too the beads observed by Mr. Baily, may have been due to irregularities in the transparency of the atmosphere.

The comparatively feeble light of the sun's disc near the limb had been already deduced by M. Bouguer from observations made by himself; and both M. Mallet and Mr. Short had alluded to the imperfect transparency of the atmosphere, as among the causes to which certain phenomena attending transits of Venus and Mercury might be referred. The suggestion of Mr. Airy may perhaps derive additional force from other considerations which Prof. A. mentions:—

1. That the appearances in question may be styled local phenomena; they having been seen by observers at certain places, but not by cotemporaneous observers at others.

2. That an imperfectly transparent medium of another kind would seem to have an effect on the visibility of these phenomena:—Mr. Paine having remarked, that through a double screen, composed of one light-red and one light-green glass, which he employed in observing the partial eclipses of 1832 and 1836, and those which were central in 1834 and 1838, "no one of the irregularities described by Mr. Baily has ever been perceived"; though observers in his vicinity through other screens saw in 1834 many or most of the usual phenomena, and they have frequently been seen through screens of other colors than red.

3. M. de la Hire observed, that when the sun was artificially eclipsed by an unpolished globe of stone, the interior of the ring

that was formed around it was broken and uneven. Prof. A. remarks, however, that it is not stated whether any portion of the actual disc was exposed in M. de la Hire's experiment; and that the appearance may perhaps admit of some other explanation.

Yet, Prof. A finds it difficult to apply Mr. Airy's suggestion to the phenomena immediately in question, because:—

1. Although the light received from the central portion of the sun's disc may be, and very probably is, characteristically different from that of the limb, yet this difference does appear to be one of *intensity*. For when experiments are made with a photometer, constructed with reference to the properties of polarized light, in which are employed two partially or entirely superimposed images, the light produced at the points of coincidence is found to be perfectly white; from which it follows, as has been remarked by M. Arago, 1. that the light of the parts near the limb is as intense as that of the center; 2. that the colors of the images are complementary to each of the other. Professor Bache's experiments at Philadelphia during the solar eclipse of 1831, with a Leslie's photometer, also point to the conclusion, that the heat at least, if not the light, which proceeded from the region near the edge of the disc, was of the average intensity.

2. The phenomena in question do not seem to have been visible to Mr. Airy under the very circumstances that suggested his explanation; though before the sun finally disappeared, he "saw the narrow ring" of his disc "not quite so bright as before."

3. The relatively faint "umbræ" or shallows, which usually surround the solar spots, do not exhibit a broken outline, though they continue to be visible, when near the sun's limb, and the limb is very seldom indented as they enter upon or leave the disc; facts not reconcilable with the proposed explanation.

4. These phenomena have been observed "only in the proximity of the borders of the sun and moon."

5. The progress of these phenomena seems to be not independent of the moon's position, since in 1836 especially, as the moon advanced, the dark spaces became narrower as well as longer. This might be referred to increased irradiation, as more of the sun was then exposed: but—

6. The scars or indentations of the moon's disc, produced by the apparent intrusion of portions of the sun at the formation of the drops, are not increased at the formation of the annulus; but on the contrary, the thread of light is not as wide as the drops that precede it.

7. Similar and seemingly analogous phenomena have been observed during the transits of Venus; and this, not only when the sun was near the horizon, and his limb somewhat unsteady, but when he was near the zenith, his limb perfectly defined and free from tremulousness, and the air remarkably serene and clear.

8. The progress of these analogous phenomena seems to have been not independent of the motion and position of the planet; as is shown by the first nine paragraphs of Table IV.

These considerations, which Prof. A. amplifies and supports by numerous references, lead him to the conclusion, that the imperfect transparency of the atmosphere, though not inappropriate, and though perhaps under certain circumstances not inefficient, is not *adequate* to the production and explanation of these phenomena.

He rejects for the same reasons the supposition that they are due to irregularities of terrestrial refraction; and considering the notion that they have their origin in peculiarities of vision or the imperfections of telescopes, as fully negated by the number and character of the observers, and the variety and excellence of their instruments, he determines, 1, that the causes of the phenomena must be sought for beyond the limits of the earth's atmosphere, and 2. that, as no efficient intervening medium is known to exist, they must reside either in the sun or in the moon.

He then gives a summary of the reasonings, which would refer them to the action of a solar envelope; 1. that it is at the sun's edge, (where of course such an envelope should have the greatest effect) that the phenomena have been seen almost exclusively, and 2. that when similar appearances have been observed at the transits of Venus and Mercury, the planet has also been, almost without an exception, at or near the edge of the sun's disc. But he regards these, as opposed successfully, 1. by arguments founded on the photometrical observations already referred to: and by the additional considerations, 2. that the phenomena are not visible alike at all places where the eclipse is visible, though circumstances be otherwise favorable: 3. that the appropriate effect of such an envelope is not observed on the umbræ of the solar spots: 4. that should such envelope be atmospheric, there would probably be a displacement of the solar spots by refraction as they recede from the center; and no such change has been observed: 5. that the hypothesis leaves unexplained the diminished depth of the indentations of the moon's limb in the formation of the complete annulus.

The 2d and 5th, and possibly also the 3d of these objections apply equally to the somewhat analogous supposition, that the phenomena are due to a characteristic difference in the properties of the solar rays, proceeding from the outer and from the central portions of the disc. Prof. A. adverts incidentally, and somewhat at large, under this head, to the well-devised and highly successful observations of M. Bessel on the eclipse of May 1836, and afterwards; by which he has established the existence of a very, bright envelope of moderate extent about the sun. He refers too, to the conjecture of Herr Beer in 1833, that there was such an envelope, and to the "ribbon of light" observed by M. de Ferrer in 1806, and similar appearances witnessed by himself in 1834, as also indicating its existence. But he does not regard these views, as having sufficient claims to explain the phenomena under consideration; especially in view of the fifth reason exhibited in the preceding paragraph.

Prof. A. next reviews several other hypotheses, which have been offered in explanation of the phenomena of Table II, especially of those which form paragraphs 1 and 4. Most of these, he remarks, have been well disposed of by Sir James South. Two of them, implying illusion on the part of the observer, or defect in his instrument, have been already noticed while discussing the phenomena of Table I: Sir James South speaks of it as remarkable, in reference to the latter of these suggestions, that the *only* case of anomaly, contributed by the Royal Observatory at Greenwich, was observed through a telescope with a forty-six inch triple-object glass, which has usually been supposed to show a remarkably small image of a star. The third hypothesis,—that of a lunar atmosphere,—Professor A. hesitates to admit; and he quotes in regard to it the remark of Sir James South, that "if such an atmosphere do exist, its effects must be the same upon all stars of a similar color," and should be expected "at every occultation which occurred": yet the instances are "almost infinitely few," in which a star undergoes either "derangement of position, diminution of splendour, or change of colour," at the moon's approach. He gives the argument on this point in the words of Sir James; noticing, however, by a reference to paragraphs 1 and 2 of Table II, the inaccuracy of Sir James's statement, that "occultations have never been preceded by a distortion of the planet," and that "an overlapping of the planet's limb on that of the moon is as yet an unknown phenomenon"; and on the whole, he appears to adopt the conclusion of that astronomer, that "amongst the immense mass of observed occultations, the existence of a lunar atmosphere, even of extreme tenuity, is so feebly supported," that it cannot be considered as adequately proved. The fourth hypothesis, that of irradiation, he rejects, with Sir James South, because instances occur sometimes of projection on the dark disc; as on the 9th Sept. 1718, when according to Maraldi and others, the immersion of a *small* star took place about the very middle of a total eclipse of the moon, and yet an adhesion to the edge of the disc was seen for several seconds. So also the fifth, viz. the "difference in refrangibility of the rays from the moon and those from the star, arising from their difference of colour"; because white stars have been, apparently projected on the disc as well as red, and Mars, though peculiarly red, has not been so projected. The sixth hypothesis, that of M. de Mairan, supposes a lunar atmosphere less dense than the ether in which "this planet" floats, and attributes the phenomena of projection to a negative diffraction or inflexion:—this may be characterized rather as a conjecture than an opinion, and is not sustained by known facts.

The conclusion to which Prof. Alexander arrives upon this review, is that in the present state of our knowledge we are not prepared to assign the physical cause or causes of the phenomena in question. He thinks, however, that many of those which have been arranged in his tables, exhibit a seeming conformity to two distinct laws, which *govern*, if they do not *regulate*, the action of the moon upon the light which passes near her surface. He expresses these laws in the following terms:—

LAW I. On the near approach of the moon, a minute beam of light is, in some cases, bent into a physical curve, the direction of which is *concave* toward the surface of the moon, in the region in which the beam escapes from her influence. As the moon advances, this flexure within certain limits increases; and when it has arrived at its maximum, the beam is cut off: the process being inverted in the case of an emerging beam.

LAW II. In other cases, or in the neighborhood of other portions of the moon's limb, a minute beam of light is sometimes bent into a physical curve, the direction of which is convex toward the surface of the moon, in the region in which the beam escapes from her influence. As the moon advances, this flexure within certain limits increases; and when it has arrived at its maximum, the beam is cut off: the process being inverted in the case of an emerging beam.

For the sake of brevity, Prof. A. speaks of this action, when the curve is concave toward the moon, as a *distortion outward*,—when convex, as a *distortion inward*—and for the two, he sometimes uses the phrase the *distorting action of the moon*. The distortion outward is represented on the next page, by figure 1, and the distortion inward by figure 2; both being of course enormously exaggerated in these representations.

Professor A. further qualifies the laws thus expressed, by observing that each is *local* and *specific* in its operation: local, because the limits of its action will be found at intervals along that edge of the disc, near which the beam of light passes: specific, because scarce any other light seems to be decidedly subject to it except such as proceeds from the sun's disc or its immediate neighborhood, and from certain stars.

He then proceeds to show how the laws thus qualified explain and reconcile many of the observed phenomena.

I. If a minute beam from a star, which is about to he occulted, pass near to that region of the moon's limb at which the distortion outward may prevail, the rays of light will be more and more bent as in fig. 1, until the flexure arrives at its maximum, the moon's edge having advanced to M, or very near it. During this interval, the spectator at E will see the star in the successive directions of the straight portions of the lines E1, E2, &c.; i.e. he will continually see the star very near the moon's edge, or—if it have a small spurious

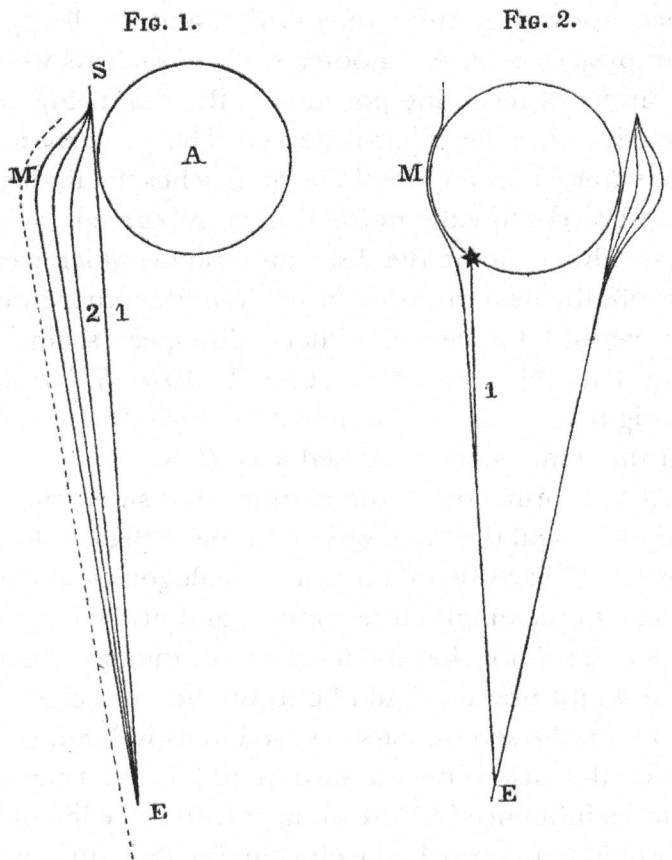

disc—in apparent contact with the edge, and seeming therefore to adhere to the limb. As the moon advances still further, the ray which passes by M will be cut off; the next distorted beam, being bent in a direction sensibly parallel to the last, as represented by the dotted line in the figure, can no longer reach E; so the edge of the moon having arrived at or near to M, the star will suddenly disappear. This is explained phenomenon 4 *of Tab. II.* The same species of action will also account for the similar appearances which occur in an inverse order, at the emersion whether of sun or star; including thus the principal phenomena described in *Tab. I.* 29.

II. If the minute beam from the star which is about to be occulted pass near to that region of the moon at which the distortion inward may prevail, then will the rays of light which the moon approaches be successively more and more bent, until the edge of the moon arrive at or near to M in fig. 2, when the star will be seen by the spectator at E in the direction E 1, and it will appear as in the figure, projected on the moon's surface; the rays which reach his eye from the intervening portions of the disc being either not thus affected, or in a far inferior degree. This projection must for like reasons commence before the edge reaches the limit at M; and as the rays are continually more and more curved, the star will continue to advance upon the disc, until the ray which passes by M will be cut off: the next distorted beam, as in the other case, cannot reach the eye; and the star will suddenly disappear, seeming to leap out of sight. Thus the phenomenon described *Tab. II.* 1, is accounted for in its origin, progress, and termination; including also the rapid motion at the immersion, described *Tab. II.* 8.

III. At the formation of the annulus, the sun's limb which is about to appear, and the luminous envelope of Bessel which closely surrounds it, will each be in a measure analogous to a continuous curve of stars in seemingly close contact, and just about to emerge from the moon's limb. Now the tendency of the distortion inward, wherever it might prevail, would be in the first instance to project upon the moon the several stars exposed to its influence; while the tendency of the distortion outward would be to keep the stars exposed to its influence in close contact with the edge of the disc; and both would most probably be efficient for about the same length of time.*

It might be presumed, therefore, if both species of action should exist at intervals along the visibly continuous arch of the sun's limb, or its luminous envelope, that several isolated spots of light would be seen indenting the edge of the moon's limb, if not actually projected on it; while at other places along the edge of the disc, portions of the luminous arch would scarcely be seen at all, being kept in close contact with the edge by the other species of distortion. As, however, the light of the luminous envelope has been observed to diminish in intensity in a direction from the sun, it might be that the light subject to the action of the distortion inward,

---

* This is ascertained by a comparison of the mean durations of the projection and adhesion of stars, phenomena 1 and 4 of Tab. II.

would not all be affected to the same extent, and thus no isolated spot be seen projected on the disc, though the limb must still be indented. The appearance arising from both actions combined would therefore resemble that of a row of drops, connected by slender filaments. But should the distortion inward precede the other a very little, the indentations would first be seen alone, and present the appearance of a row of drops with intervening dark spaces. As the moon afterwards advanced, the drops, in accordance with Law II. would begin to pass off; while the light, having slightly emerged at the intervening spots along the edge, would continue to adhere in the form of slender threads, or broader bands, connecting the drops. At a later period the drops would pass off, the intervening threads would be released from their adhesion, and the whole would run together, completing a comparatively smooth ring, whose breadth might at first be less than that of the drops.

The same, in an inverse order, might take place at the dissolution of the annulus, which implies the immersion of a luminous arch. The emersion of a small portion of the sun from a total eclipse, would moreover be analogous to the formation of the annulus, and might present like phenomena; while the immersion of the last narrow and shortened portion of the sun, at the beginning of the total obscuration, might exhibit appearances similar to those seen at the dissolution of the annulus.

All the phenomena described in the first part of *Tab. I.* 14, are thus accounted for in their commencement, progress, and termination.

The exposition thus far, does not include the dark lines observed by Prof. Van Swinden in 1820, nor those, and the phenomena preceding them, which were seen in such rare perfection by Mr. Baily and others, in 1836.

Should, however, the distortion inward be very feeble, or scarcely exist at all, the indentations would at first be very small, and the portions of the luminous arch at which such distortion took place, might almost immediately pass off the disc; but the distortion outward being still efficient, the adhesion would still continue, and might either commence later or continue longer at some places than others. The bright spots must therefore appear wider and deeper, and the dark spots narrower and longer, until, the effect having reached its limit, the dark lines would break off. All this time, the ends of the dark lines toward the moon's limb would most probably be separated from it by fine filaments of light, though the minuteness of these as well as the rapidity of the other changes might prevent their being observed.

The same appearances might recur in an inverse order, at the dissolution of the annulus, and thus present phenomena similar to those observed by Mr. Baily, in their origin, progress, and termination. The hypothesis explains in this manner all the phenomena of *Tab. I.* 14.

The prescribed limits of this abstract exclude the detailed applications of Prof. A.'s law to the numerous other phenomena which it reconciles. In general, it may be said, that among these phenomena are, the elongation, shortening, and flexure of cusps, *Tab. I.* 4, and *III.* 1,—the apparent gliding of a star on the edge of the disc before immersion, *Tab. II.* 5,—the distortion of the moon's disc, *Tab. I.* 16,—the division of the planet's disc, *Tab. II.* 3,—the prominent dark points on the sun's disc at the commencement of an eclipse, *Tab. I.* 2 *and* 29,—the dark lines on the sun's limb, near the beginning or ending of the eclipse, *Tab. I.* 3,—the illuminated solar spot, *Tab. I.* 8,—the sudden formation of the annulus, *Tab. 1.* 17,—the brilliant spot on the moon's disc after the total immersion of the sun, *Tab. I.* 25.

Prof. A. remarks, in continuation, that the hypothesis of the distorting action of the moon is confirmed also by the fact, that the mean duration of the drops and dark lines in eclipses, so far as ascertained, is nearly identical with that of the projections and adhesions of stars in occultations;—the duration of the drops having been in 1806, $4^s.5$ —their mean duration in 1820, $2^s.44$, and in 1836, $9^s.0$—and their duration in 1838 at the beginning and end, $6^s.33$, and at the annular eclipse, $4^s.7$; giving a general mean of $5^s.2$: while the mean duration of the changes at occultations is $4^s.5$,—or when two unusual instances of nearly 15^s, each are included, $5^s.0$. Further confirmation of it is found in the analogy presented by phenomena seen during transits, *Tab. IV.* 1, 3, 4, 6–13, *and* 19–24 inclusive, and *Tab. V.* 1, 2, 7, 9:—and that this analogy is real, is sustained by the appearances of Venus and Mercury, noticed in *Tab. VI.* 3–11, and *Tab. VII.* 1, 3, 4.

Traces of the local action of the moon upon light have been observed also on other occasions; *Tab. III.* 4, 5:—and effects similar to some of these distortions have been produced by artificially eclipsing the sun and stars.

Another analogy in the action of the moon and Venus upon light is presented in the tremulous motion near the moon's limb in eclipses, and the similar appearance near the limb of Venus in transits;—and that this last is due to the action of the planet appears by reference to *Tab. IV.* 1, 2, 3, 9–12, and to the fact, that the phenomena there described interfered with and quieted the unsteadiness of the sun's limb, produced by irregular refraction. The phenomena noticed in *Tab. I.* 5, may therefore be referred to the action of some unexplained cause in the moon. The flashes, described *Tab. I.* 6, seem to be dependent upon the action of some cause in the moon, since they usually come from the moon's edge; and the same may be said of the spots and traces of light, mentioned in, *Tab. I.* 9.

Prof. A. next considers the *corona, Tab. I.* 20, and examines in succession the hypotheses to which it has given rise. After discussing those which refer it to the action of the earth's atmosphere, and to the diffractive influence of the moon, and giving the arguments at large in favor of a lunar atmosphere, he rejects all these, as either contrary to known facts, or inadequate to explain the phenomenon; and passes to the inquiry, whether it may not be due to the same widely diffused luminous substance, which is also apparent in the zodiacal light. He presents in detail the considerations on both sides of this question; and concludes, that notwithstanding the measurement made by the associates of M. Arago at Perpignan in 1842, which led them to the conclusion that the corona was concentric with the moon, there is yet reason to accept this hypothesis, as the one most generally accordant with all the facts that have been observed.

He then suggests distinct explanations for the several other phenomena mentioned in his tables; referring some of them to atmospheric action, some to earth-light, others to the effects of contrast, others again to a glare of light within the telescope, &c. &c., and closes his elaborate paper by some conjectures as to the cause of the distinct outline exhibited by the ribbon of light, 26 *of Tab. I.*, in 1806 and 1834; as well as that of the distortion, of which he has endeavored to trace the laws and mode of action.

The great heat of the sun's surface, he says, would seem to render it possible that a comparative vacuum may be maintained there; or in other words, that the density of his envelope may be almost incomparably less at his surface than at a very short distance from it. A somewhat abrupt change of density will thus be effected; and the light which arrives at the limit of this vacuum will be copiously reflected, upon the principle that light is always thus reflected at the common surface of media differing in density. This light will pass through the nebulosity on all sides, as the sun's rays shine through a comet; and the limit of the vacuum being seen very obliquely or foreshortened near the edges, it must present something like a distinct boundary.

Again; he proceeds, whatever may be said of the reason assigned by Dr. Halley for the dim light of the sun before his immersion; namely, the vapors in the moon's atmosphere, collected during the long night; yet it would seem to be very certain, that the temperature of the western edge, which is just then passing from night to day, must be decidedly lower than that of the eastern edge, which has been exposed to the sun's light for many days together. The same will, vice versa, be true with regard to the other edge in its turn.*

The distribution of temperature would, on other accounts also, be very unequal over that part of the surface,—somewhat more than a hemisphere,—which alone is ever visible to us. For the great annular pits would tend to concentrate the heat reflected from their surfaces; and should there be any thing vaporable near their centers, it must feel the effect of this concentrated and long continued heat. A distribution of unequally heated spots would thus be produced; which spots, moreover, would again in all probability be unequally

---

* See Sir John Herschell's Treatise on Astronomy, page 219. VOL. III.—2 D

cooled during the long night. The local action upon light, as it passed near these heated pits, might therefore be in some respects the negative of that exerted by the colder portions, and none would probably be colder than those mountains which have not the crater-like form.

In these disposing causes, he says, may perhaps be found the origin of the two species of distortion which have been described in this paper. If, moreover, the distortion inward belong to the pits, and the distortion outward to the eminences; then, at the formation of the annulus, the former would a little precede the latter,—a supposition which, as has been seen, explains the phenomena in some cases more fully.

Apart from any such speculations, however, the following, he remarks, is a summary of the conclusions to which the preceding investigation appears to lead:—

1. That if the moon were surrounded by an atmospheric envelope, such envelope must be affected by a terrestrial and anti-terrestrial tide, the great waves of which must uniformly be accumulated nearly above the center of the disc, and the point diametrically opposite, thus leaving the edges comparatively destitute.

2. That among the causes of the physical phenomena attending solar eclipses and occultations;—besides atmospheric action, earth-light, and others;—must be specially recognized a distorting action, of a two-fold character, exerted upon light by the moon: —that this action is—1. local; 2. specific; and that it seems to be exerted either upon the light at the sun's limb, or upon a luminous substance or substances surrounding the sun:—and

3. That to this luminous substance is due the corona.

Prof. A. adds some supplementary remarks, in the course of which he quotes several other observations, as furnishing additional proof of the existence of specific peculiarities in envelopes of the various planets of the solar system. These, he says, seem to afford ground for a somewhat broad generalization, viz :—*that an atmosphere properly so called is probably peculiar to the earth.*

Professor Bache, from the sub-committee of arrangements, announced that the program of the meeting had now been gone through.

In the absence of the President from indisposition, Dr. Chapman, one of the Vice-Presidents, then made some parting remarks on the manner in which the hundredth anniversary of the Society had been celebrated. He adverted to the dignified and imposing assemblage which had done honor to the anniversary ceremonies, the high scientific character of the gentlemen who had taken part in the special meeting which was now about to close, the ability and enlarged scope of the communications presented before it, and the intelligent interest manifested in all its proceedings by the numerous strangers and citizens who had witnessed them.

Dr. C. then spoke of the agreeable recollections which would remain after this celebration, the incitement it had given to fresh scientific labor, the sympathies it had awakened and revived among men engaged in various though kindred pursuits, and the personal friendships it had so happily established. He tendered the thanks of the Society to all who had honored it by their presence on the occasion; and bidding them a respectful farewell, expressed an earnest wish that the recurrence of so grateful a communion of mind might not be long postponed.

Note to Mr. Tyson's Paper, Page 124.

In a note to the Reporter, Mr. Tyson refers to an original document in the possession of Mr. George M. Justice, of Philadelphia, and recently communicated by him to the Historical Society of Pennsylvania, as correcting the statement, which Mr. Bancroft and others have admitted, that William Penn was at one time a slaveholder.

APPENDIX.

REPORT OF THE COMMITTEE OF ARRANGEMENTS,

Presented June 16, 1843,

AND

RESOLUTIONS ADOPTED THEREUPON

BY THE SOCIETY.

THE Committee appointed to make arrangements for the Celebration of the Hundredth Anniversary of the American Philosophical Society, respectfully report: that although from time to time they have submitted to the Society the arrangements in progress, with a view to obtain instructions and authority in regard to them; yet having completed nearly all the duties growing out of the occasion, they deem it proper to put upon record, in a single report, an outline of the entire proceedings, and to call the attention of the Society to the unfinished business of the meetings succeeding the Anniversary.

On the report of a Committee of Inquiry* into the expediency of celebrating the hundredth Anniversary, the Society determined on the 3d of March last, to hold the celebration, to issue a special summons to the members requesting their attendance, and to invite one of the members to deliver an address on the 25th day of May, upon the history and prospects of the Society. Dr. R. M. Patterson, one of the Vice Presidents, was by the unanimous vote of the Society requested to prepare the address; and the undersigned were appointed a committee to carry into effect the wishes of the Society in the arrangements for the celebration. Dr. Patterson acceded to

---

* Consisting of Dr. Chapman, Mr. Kane, Dr. Dunglison, Dr. Ludlow and Mr. Fraley.

241

the request of the Society; and a circular note to members and societies in correspondence with us was at once issued, informing them of the intended celebration. The Committee also divided themselves into subcommittees, who took charge of different branches of the arrangements.

In the progress of the discussions of the Committee, and of their conferences with members of the Society, the conviction grew that this occasion was a fitting one, not only to review the past, tracing the Society from its birth, and presenting it as the guardian of the scientific reputation of its founders, but also to mark the present condition of science in the United States, by calling its cultivators to meetings where they might bring their latest contributions in the walks which they had chosen. The lateness of the period at which this plan was determined on rendered success very doubtful, and prevented as general an appeal to the scientific men of our country as was desirable. A special circular was, however, addressed to many, who it was deemed probable might attend in person, or send papers to be read on the occasion.

This call of the Society was responded to in the most gratifying manner. The letters, generally, from individuals and societies, breathed a spirit of earnest zeal in the cause of science, and of kind regard for the welfare of the oldest scientific association of the country. Many of the writers were eloquent in expressing their sense of the interest of the occasion, of the importance of holding up to view the labors of the pioneers of science, and of assembling around our Society the men of science of the country, for mutual encouragement, support, and improvement.

The tone of these letters greatly increased the interest felt in the preparations. The circumstances of the Society not permitting calls upon its funds, the members cheerfully came forward with contributions to defray the expenses of the meetings, and especially to secure the immediate publication in the Proceedings of abstracts or notices of the papers which might be read. Announcements of intention to be present, and of communications prepared or in progress, were daily received; and it became plain, that the efforts of the members resident in the city to give interest to the occasion, would be met with a kindred spirit by those in other places, and by our correspondents.

The oration delivered at the Musical Fund Hall, on the 25th of May, was attended by delegates from some of the societies in correspondence with us, by many invited guests from other parts of the country, by different associations of our city, the members of the religious bodies then in session, the judicial and municipal authorities, the professors and students of the University, Colleges, and High School, the officers of the Army and Navy stationed at Philadelphia, and by a respectable number of our citizens. One of the venerable seniors of the Society, the Rev. Dr. Ashbel Green, was also present on the occasion. The ceremonies were opened by a few felicitous remarks from our President, and a prayer by the Rev. Provost Ludlow, and the Rev. Dr. Dorr was invited to close them with an appropriate benediction. The thanks of the Society will doubtless be presented to the orator of the day, Vice President R. M. Patterson, for his most interesting account of the early history and progress of the Society, and the Committee propose that a copy of the address, and the remarks of Mr. Du Ponceau, and the prayer of Dr. Ludlow, be asked for publication in the Proceedings.

The meetings for the transaction of scientific business, which it was originally expected would terminate on Friday, were not only held on that day, but were from necessity adjourned over to the Saturday, Monday, and Tuesday, following the Anniversary; the morning session usually occupying from 10 o'clock, A.M. until 2 P.M., and the evening session from $7 \frac{1}{2}$ until $10 \frac{1}{2}$, P.M. At the opening of Friday's meeting there were already thirty-two papers on the list, and the number increased before the following Tuesday to forty-five.

The names of the authors sufficiently assured the excellence of their communications, and the subjects generally were of high interest. Twenty-one papers were presented upon Physical Science and Chemistry, ten upon Natural History, Geology, Geography, and Ethnography, four upon Mechanics, four upon Medical Science, three upon Mathematics, and three upon History and Biography.

The interest of the meetings was much increased by the presence, generally, of the authors of memoirs, and by the care taken in preparing suitable diagrams and drawings for illustration. In this connection, the Committee have pleasure in referring to the kind services of certain pupils of the High School, detailed with the consent of their parents by the Principal, to assist in preparing drawings, and to attend at the meetings as aids to the Committee of Arrangement. Many of the diagrams were executed by these youths in a style which did them great credit; and to their faithful assiduity during the meetings, and at other times, the Committee were much indebted. The attendance at all the meetings of many strangers and citizens, not members of the Society, manifesting untiring attention in the proceedings, an intelligent zeal in behalf of science, and a warm interest in the success of its cultivators, was in the highest degree cheering.

The meetings were closed on Tuesday evening by an address from Vice President Chapman; and the members and their guests parted with feelings of mutual regret that their happy reunion had terminated, and with most agreeable, and we trust profitable, recollections of the occasion. The impression made upon the scientific gentleman from a distance found vent in the hope unanimously expressed that this would be but the beginning of a series of meetings under the auspices of the Society. The Committee would suggest that the thanks of the Society be returned to those associations who sent delegates, and to the gentlemen who attended the meetings.

In the course of the preparations for the Anniversary, many members of the Society were forcibly struck with the change which has gradually grown up in our list of members, from names embracing the talent and influence of the different parts of our own Commonwealth and of the United States, to those included in more restricted and local influences. The cause of this does not fully appear: Exclusiveness has certainly never been part of the system of the Society. Even many of the leading scientific men in different sections of our country are not enrolled among our members; and the Society owes the cooperation, which has so signally marked its late Anniversary, more to the wide views and generous spirit of the cultivators of science, than to the care taken to propitiate them. A suggestion upon this head, the Committee trust will find the favor, when presented to the Society as a body, which it has already very generally met from the members, when presented to them individually.

Toward the close of the meetings, a resolution was laid upon the table by one of our members from a neighboring State, which the Committee deem of high importance, and which they recommend to the most favorable consideration of the Society. A meeting such as the resolution proposes, will enable the Society to perform a duty too long postponed. It will indeed have but partially fulfilled its duty to its early members, if it fails to make opportunity to bring their labors prominently forward, by a careful and elaborate review of them. Already misconception, in regard to the nature and extent of Franklin's discoveries in electricity, exists, and is obviously increasing:—the credit of the optical researches of Rittenhouse is awarded to others. Now is the time to correct these and other errors. A remarkably liberal and truth-seeking spirit is abroad among scientific men, less disturbed by local and national jealousies now than perhaps at any previous time. It invites us to bring in our contributions to the history of science, and to vindicate the claims of Americans to discoveries which otherwise may be lost to them. Almost every learned society has some provision for an occasional review of the labors of its members; and in our own this review should, by law, be made annually. The task thus imposed is indeed one of peculiar delicacy; for the works of the living, in all the freshness of recent production, are to be criticized. But no such difficulty is presented by the labors of a past century: distance has sufficiently enlarged the field of view, to give to them their true positions and relative eminence; while it has not yet obscured and made them indistinct. Besides, we owe it as a debt to those who have gone before us, and whose name we inherit. It is the appropriate memorial of filial piety.

In accordance with the views presented in the foregoing Report the Committee offer for the consideration of the Society the annexed resolutions.

N. CHAPMAN,
J. K. KANE,
ROBLEY DUNGLISON,
JOHN LUDLOW,
F. FRALEY,
A. D. BACHE,
GEO. W. SMITH,
Committee.

1. Resolved, That the thanks of the Society be presented to the several learned societies and individuals, who contributed by their presence and cooperation to the interest and value of the meetings, which followed the late Anniversary.

2. Resolved, That the thanks of the Society be tendered to Vice-President Patterson, for the interesting, able, and instructive Discourse, pronounced by him at the centenary Anniversary of the Society, and that it be published under the direction of the Committee of Arrangements.

3. Resolved, That copies be also requested for publication, of the appropriate address delivered by the President of the Society, and of the eloquent prayer of the Reverend Provost Ludlow at the centenary celebration.

4. Resolved, That the papers, presented at the special meeting, and which may be considered as intended by their authors for publication in the Transactions, be referred to Committees, under the 1st section of chapter X. of the Laws of the Society.

5. Resolved, That so much of the resolution submitted by Prof. Henry at the special meeting, as relates to the celebration of future anniversaries, be referred to a Committee of five, to consider and report thereon.

OFFICERS

OF THE

AMERICAN PHILOSOPHICAL SOCIETY,

MAY 25, 1843.

President.
Peter S. Du Ponceau, LL.D.

Vice-Presidents.
Nathaniel Chapman, M.D.
Robert M. Patterson, M.D.
Franklin Bache, M.D.

Secretaries.
John K. Kane,
Alexander Dallas Bache, LL.D.
Robley Dunglison, M.D.
Joshua Francis Fisher.

Curators.
John Price Wetherill,
Isaac Hays, M.D.
Franklin Peale.

Treasurer.
George Ord.

Counsellors.
Robert Hare, M.D.
William Hembel,
Charles D. Meigs, M.D.
Henry Vethake, LL.D.
Clement C. Biddle,
William Short,
Joseph Henry, LL.D.
Benjamin Dorr, D.D.

Thomas Biddle,
Gouverneur Emerson, M.D.
Isaac Lea,
Hartman Kuhn.

MEMBERS

OF THE

AMERICAN PHILOSOPHICAL SOCIETY,

MAY 25, 1843,

With their Original Numbers, and the Dates of their Election.

454.	Benjamin Chew, of Pennsylvania,	19 Jan.	1787
504.	Ashbel Green, D.D., of Pennsylvania,	16 Jan.	1789
529.	James Ross, of Pennsylvania,	21 Jan.	1791
532.	Albert Gallatin, of Pennsylvania,	21 Jan.	1791
551.	Peter Stephen Du Ponceau, of Pennsylvania,	15 July	1791
559.	Anthony Renatus C. M. de la Forest,	20 Jan.	1792
570.	John Trumbull, of Connecticut,	20 July	1792
610.	A. I. Larocque, of France,	15 April	1796
618.	Charles Caldwell, M.D., of Kentucky,	21 Oct.	1796
624.	Samuel Harrison Smith, of the District of Columbia,	20 Jan.	1797
630.	John Guillemard, of Oxford, England,	21 July	1797
652.	Samuel Miller, D.D., of New Jersey,	18 April	1800
670.	William Stephen Jacobs, M.D., of St. Croix,	16 July	1802
672.	James Mease, M.D., of Pennsylvania,	16 July	1802
677.	Robert Hare, M.D., of Pennsylvania,	21 Jan.	1803
686.	Robert Gilmor, of Maryland,	21 Oct.	1803
690.	Manuel Godoy, Prince of Peace,	20 April	1804
691.	Pedro Cevallos, of Spain,	20 April	1804
694.	William Short, of Virginia,	20 July	1804
695.	Baron Alexander Von Humboldt, of Prussia,	20 July	1804
704.	Samuel Moore, M.D., of Pennsylvania,	18 Jan.	1805
706.	Benjamin Silliman, M.D., of Connecticut,	18 Jan.	1805
712.	Joseph Cloud, of Pennsylvania,	17 Jan.	1806
713.	Samuel B. Wylie, D.D., of Pennsylvania,	17 Jan.	1806

719.	Mahlon Dickerson, of New Jersey,	16 Jan.	1807
721.	Nathaniel Chapman, M.D., of Pennsylvania,	17 April	1807
723.	Ferdinand Rudolph Hassler, of the District of Columbia,	17 April	1807
726.	James Gibson, of Pennsylvania,	17 April	1807
728.	Charles Philibert de l'Asteyrie, of France,	16 Oct.	1807
730.	Horace Binney, of Pennsylvania,	15 July	1808
733.	Ross Cuthbert, of Lower Canada,	20 Jan.	1809
736.	George William Featherstonhaugh, of New York,	21 April	1809
737.	David B. Warden, of Paris,	21 April	1809
738.	Robert M. Patterson, M.D., of Pennsylvania,	21 April	1809
742.	F. André Michaux, of France,	21 April	1809
752.	John Davis, of Massachusetts,	18 Jan.	1811
753.	Charles J. Wister, of Pennsylvania,	18 Jan.	1811
755.	Robert Walsh, of Pennsylvania,	17 Jan.	1812
757.	*Robert Adrain, of New Jersey,	17 July	1812
761.	Constant Dumeril, of Paris,	16 April	1813
763.	John Sergeant, of Pennsylvania,	16 April	1813
764.	Nicholas Biddle, of Pennsylvania,	16 April	1813
765.	William P. C. Barton, M.D., of Pennsylvania,	16 Oct.	1813
766.	William Meredith, of Pennsylvania,	16 Oct.	1813
767.	Charles Chauncey, of Pennsylvania,	16 Oct.	1813
769.	William Hembel, of Pennsylvania,	16 Oct.	1813
773.	Frederick Beasley, D.D., of New Jersey,	21 Jan.	1814
775.	Gen. Joseph Gardiner Swift,	15 April	1814
776.	Thomas Gilpin, of Pennsylvania,	15 April	1814
778.	John Gummere, of New Jersey,	15 July	1814
782.	John Morin Scott, of Pennsylvania,	21 April	1815
783.	Joseph Hartshorne, M.D., of Pennsylvania,	21 April	1815
785.	Charles Jared Ingersoll, of Pennsylvania,	21 April	1815
790.	Gerhard Troost, M.D., of Tennessee,	19 Jan.	1816
791.	Joseph Reed, of Pennsylvania,	19 Jan.	1816
797.	Charles Alexander Le Sueur, of Paris,	17 Jan.	1817
800.	John C. Otto, M.D., of Pennsylvania,	17 Jan.	1817
801.	Richard Rush, of Pennsylvania,	17 Jan.	1817
805.	Joseph Von Hammer, of Vienna,	18 July	1817

---

* Died 10 Aug. 1843.

806.	William Gaston, of North Carolina,	18 July	1817
807.	Charles Fenton Mercer, of Virginia,	18 July	1817
809.	Eugenius Nulty, of Pennsylvania,	17 Oct.	1817
811.	George Ord, of Pennsylvania,	17 Oct.	1817
812.	Thomas Nuttall, of England,	17 Oct.	1817
816.	John Quincy Adams, of Massachusetts,	17 April	1818
819.	James G. Thomson, of Pennsylvania,	17 April	1818
820.	Parker Cleaveland, of Maine,	17 April	1818
821.	John C. Warren, M.D., of Massachusetts,	17 April	1818
822.	James Jackson, M.D., of Massachusetts,	17 April	1818
824.	Gottlieb Fischer, of Russia,	17 April	1818
825.	Daniel Drake, M.D., of Ohio,	17 April	1818
826.	Jacob Bigelow, M.D., of Massachusetts,	17 April	1818
831.	H. M. Ducrotay de Blainville, of Paris,	15 Jan.	1819
833.	Guillaume Theophile Tilesius,	16 April	1819
836.	Jacob Perkins, of Massachusetts,	18 June	1819
839.	Alexander Brongniart, of Paris,	15 Oct.	1819
840.	Redmond Conyngham, of Pennsylvania,	15 Oct.	1819
843.	William E. Horner, M.D., of Pennsylvania,	15 Oct.	1819
847.	Franklin Bache, M.D., of Pennsylvania,	21 April	1820
848.	William Gibson, M.D., of Pennsylvania,	21 April	1820
849.	Samuel F. Jarvis, D.D., of New York,	20 Oct.	1820
850.	Isaiah Lukens, of Pennsylvania,	20 Oct.	1820
851.	John Jacob Berzelius, of Sweden,	20 Oct.	1820
852.	J. A. Borgnis, of France,	20 Oct.	1820
854.	M. de Montgéry, of France,	20 Oct.	1820
855.	William Strickland, of Pennsylvania,	20 Oct.	1820
856.	John Pickering, of Massachusetts,	20 Oct.	1820
857.	Langdon Cheves, of South Carolina,	19 Jan.	1821
859.	John B. Gibson, of Pennsylvania,	19 Jan.	1821
860.	George Alexander Otis, of Massachusetts,	20 April	1821
861.	Clement C. Biddle, of Pennsylvania,	20 April	1821
864.	Peter Azelius, of Sweden,	20 April	1821
865.	Sir James Wiley, of St. Petersburg,	20 April	1821
866.	Gustavus, Count Wetterstedt, of Sweden,	20 July	1821
869.	Peter Poletica, of Russia,	18 Jan.	1822
870.	P. Pedersen, of Denmark,	18 Jan.	1822
873.	Richard Harlan, M.D., of Pennsylvania,	19 April	1822

875. Jons Svanberg, of Sweden,	19 April	1822
880. Lardner Vanuxem, of Pennsylvania,	18 Oct.	1822
884. Samuel Jackson, M.D., of Pennsylvania,	17 Jan.	1823
885. Benjamin H. Coates, M.D., of Pennsylvania,	18 April	1823
886. James Fenimore Cooper, of New York,	18 April	1823
889. Joseph, Count Survilliers,	18 April	1823
891. H. C. Schumacher, of Denmark,	18 April	1823
892. William Darlington, M.D., of Pennsylvania,	18 April	1823
893. William Bengo Collyer, of England,	18 April	1823
895. Stephen H. Long, U. S. Topographical Engineers,	17 Oct.	1823
897. Nathaniel A. Ware, of Pennsylvania,	17 Oct.	1823
899. Moses Stuart, D.D., of Massachusetts,	16 Jan.	1824
900. Henry Seybert, of Pennsylvania,	16 Jan.	1824
904. A. I. Von Krusenstern, of Russia,	16 April	1824
905. Charles Lucien Bonaparte, Prince of Canino, &c.	16 April	1824
906. Conrad I. Temminck, of Leyden,	16 July	1824
909. Count John. Laval, of Russia,	21 Jan.	1825
910. John I. Bigsby, M.D., of England,	21 Jan.	1825
911. M. Flourens, M.D., of Paris,	15 April	1825
914. John K. Kane, of Pennsylvania,	15 April	1825
916. Charles N. Bancker, of Pennsylvania,	15 April	1825
919. Joseph R. Ingersoll, of Pennsylvania,	15 July	1825
921. Philip Tidyman, M.D., of South Carolina,	21 Oct.	1825
922. Samuel Humphreys, of Pennsylvania,	20 Jan.	1826
925. Charles D. Meigs, M.D., of Pennsylvania,	21 April	1826
926. William M'Ilvaine, of Pennsylvania,	21 April	1826
927. Jacopo Graberg di Hemso, of Sweden,	21 April	1826
928. Henry De Struvé, of Russia,	20 Oct.	1826
929. Lewis Cass, of Michigan,	20 Oct.	1826
932. Joel R. Poinsett, of South Carolina,	19 Jan.	1827
933. René La Roche, M.D. of Pennsylvania,	19 Jan.	1827
934. John Price Wetherill, of Pennsylvania,	20 April	1827
935. George Emlen, of Pennsylvania,	20 April	1827
937. Marcus Bull, of Pennsylvania,	20 April	1827
939. Dr. George Maria Zecchinelli, of Padua,	20 April	1827
940. J. P. C. Cassado de Giraldes, of Portugal,	20 July	1827
942. John K. Mitchell, M.D., of Pennsylvania,	20 July	1827
944. *Noah Webster, of Connecticut,	19 Oct.	1827

* Died May 28, 1843

947.	Thomas Harris, M.D., of Pennsylvania,	18 Jan.	1828
948.	Robert Eglesfeld Griffith, M.D., of Pennsylvania,	18 Jan.	1828
950.	Samuel George Morton, M.D., of Pennsylvania,	18 Jan.	1828
951.	Adm. Jose Maria Dantes Pereira, of Portugal,	18 April	1828
952.	Henry James Anderson, M.D., of New York,	18 April	1828
953.	Isaac Lea, of Pennsylvania,	18 April	1828
954.	Samuel Betton, M.D., of Pennsylvania,	18 July	1828
955.	George Ticknor, of Massachusetts,	18 July	1828
956.	James Renwick, of New York,	17 Oct.	1828
957.	Thomas Biddle, of Pennsylvania,	16 Jan.	1829
958.	William H. De Lancey, D.D., of New York,	16 Jan.	1829
959.	Hans Christian Oersted, of Denmark,	16 Jan.	1829
960.	Baron Hyde de Neuville, of France,	16 Jan.	1829
961.	Carla Christian Rafn, of Denmark,	16 Jan.	1829
962.	Henry Wheaton, of New York,	16 Jan.	1829
963.	Alexander Dallas Bache, of Pennsylvania,	17 April	1829
965.	James Kent, of New York,	17 April	1829
966.	Josiah Quincy, of Massachusetts,	17 April	1829
967.	Washington Irving, of New York,	17 April	1829
970.	Col. J. N. B. Von Abrahamson, of Denmark,	17 July	1829
971.	George B. Wood, M.D., of Pennsylvania,	17 July	1829
973.	Francisco de Paula Quadrado, of Spain,	16 Oct.	1829
974.	M. Jomard, of Paris,	16 Oct.	1829
975.	Henry S. Tanner, of Pennsylvania,	16 Oct.	1829
976.	Daniel B. Smith, of Pennsylvania,	16 Oct.	1829
977.	Thomas Horsfield, M.D., of London,	16 Oct.	1829
980.	William Yarrel, of England,	15 Jan.	1830
982.	Jules de Wallenstein, of Russia,	15 Jan.	1830
983.	Thomas M'Euen, M.D., of Pennsylvania,	15 Jan.	1830
984.	Duke Bernard of Saxe Weimar,	16 April	1830
985.	William B. Hodgson, of Virginia,	16 April	1830
986.	Isaac Hays, M.D., of Pennsylvania,	16 April	1830
988.	William Vaughan, of London,	16 April	1830
989.	Thomas I. Wharton, of Pennsylvania,	16 July	1830
990.	Lorenzo Martini, of Turin,	15 Oct.	1830
991.	Andres del Rio, of Mexico,	15 Oct.	1830
992.	Marc Antoine Jullien, of France,	15 Oct.	1830
994.	Hyacinth Carena, of Turin,	21 Jan.	1831

995. Louis Philippe, King of the French,	21 Jan.	1831
996. Thomas P. Jones, M.D., of the District of Columbia,	21 Jan.	1831
997. Henry Vethake, of Pennsylvania,	15 April	1831
999. Edward Everett, of Massachusetts,	15 April	1831
1000. Louis M'Lane, of Delaware,	15 April	1831
1001. William C. Rives, of Virginia,	15 April	1831
1002. Alexander Everett, of Massachusetts,	15 April	1831
1003. Martin Fernandez Navarrete, of Spain,	15 July	1831
1005. John James Audubon, of New York,	15 July	1831
1006. Maj. Hartman Bache, U. S. Topographical Engineers,	21 Oct.	1831
1008. Julius D. Ducatel, M.D., of Maryland,	20 Jan.	1832
1009. Henry D. Gilpin, of Pennsylvania,	20 Jan.	1832
1011. John Bell, M.D., of Pennsylvania,	20 Jan.	1832
1012. Robley Dnnglison, M.D., of Pennsylvania,	20 Jan.	1832
1013. Steen Bille, of Denmark,	20 Jan.	1832
1014. Thomas Sergeant, of Pennsylvania,	20 Jan.	1832
1015. Theodore Lorin, of France,	20 April	1832
1016. Hugh L. Hodge, M.D., of Pennsylvania,	20 April	1832
1017. Col. John J. Abert, U. S. Topographical Engineers,	20 April	1832
1018. Juan Jose Martinez, of Spain,	20 April	1832
1020. E. S. Bring, of Sweden,	20 July	1832
1021. Professor Bujalsky, of Russia,	18 Jan.	1833
1022. Marmaduke Burrow, M.D., of Pennsylvania,	18 Jan.	1833
1023. Matthias W. Baldwin, of Pennsylvania,	18 Jan.	1833
1024. Edwin James, M.D., of New York,	18 Jan.	1833
1025. Moncure Robinson, of Virginia,	18 Jan.	1833
1026. M. J. Labouderie, of France,	19 April	1833
1027. Charles Nagy, of Hungary,	19 April	1833
1028. Jacob Randolph, M.D., of Pennsylvania,	19 April	1833
1029. Joshua Francis Fisher, of Pennsylvania,	19 April	1833
1030. Gouverneur Emerson, M.D., of Pennsylvania,	19 April	1833
1031. Henry C. Carey, of Pennsylvania,	19 April	1833
1032. Henry R. Schoolcraft, of Michigan,	19 July	1833
1033. Viscount Santarem, of Portugal,	19 July	1833
1034. Titian R. Peale, of Pennsylvania,	19 July	1833
1035. Franklin Peale, of Pennsylvania,	18 Oct.	1833
1036. Samuel Vaughan Merrick, of Pennsylvania,	18 Oct.	1833

1037. Henry J. Williams, of Pennsylvania,	18 Oct.	1833
1038. Henry D. Rogers, of Pennsylvania,	2 Jan.	1835
1039. James P. Espy, of Pennsylvania,	2 Jan.	1835
1040. Edward H. Courtenay, of Pennsylvania,	2 Jan.	1835
1041. Charles W. Short, M.D., of Kentucky,	2 Jan.	1835
1042. John Brockenbrough, of Virginia,	2 Jan.	1835
1044. John Torrey, M.D., of New York,	2 Jan.	1835
1045. Joseph Henry, of New Jersey,	2 Jan.	1835
1046. D. Francis Condie, M.D., of Pennsylvania,	2 Jan.	1835
1048. William B. Rogers, of Virginia,	17 July	1835
1049. Thomas Sully, of Pennsylvania,	17 July	1835
1050. Charles A. Agardh, of Sweden,	17 July	1835
1051. C. C. Von Leonhard, of Heidelberg,	15 Jan.	1836
1052. C. G. C. Reinwardt, of Leyden,	15 Jan.	1836
1053. D. Manuel Naxera, of Mexico,	15 Jan.	1836
1054. Chevalier Morelli, of Naples,	15 Jan.	1836
1055. Job R. Tyson, of Pennsylvania,	15 Jan.	1836
1056. Nathan Dunn, of Pennsylvania,	15 Jan.	1836
1057. John Griscom, of Pennsylvania,	15 Jan.	1836
1058. J. S. Da Costa Macedo, of Portugal,	15 April	1836
1059. Nicholas Carlisle, of London,	15 April	1836
1060. Granville Penn, of England,	15 April	1836
1061. Col. Joseph G. Totten, U. S. Engineers,	21 Oct.	1836
1062. M. Roux de Rochelle, of France,	21 Oct.	1836
1063. Dr. Mariano Galvez, of Guatemala,	21 Oct.	1836
1065. George Campbell, of Pennsylvania,	21 April	1837
1066. John Green Crosse, of England,	21 April	1837
1067. Jared Sparks, of Massachusetts,	21 April	1837
1068. Charles R. Leslie, of London,	21 April	1837
1069. James Cowles Pritchard, M.D., of England,	. 21 April	1837
1071. George Tucker, of Virginia,	21 April	1837
1072. William Jenks, D.D., of Massachusetts,	21 July	1837
1073. Sears C. Walker, of Pennsylvania,	20 Oct.	1837
1074. Joseph Saxton, of Pennsylvania,	20 Oct.	1837
1075. William Morris Meredith, of Pennsylvania,	20 Oct.	1837
1076. Thomas Dunlap, of Pennsylvania,	20 Oct.	1837
1077. Daniel Webster, of Massachusetts,	20 Oct.	1837
1078. Capt. Andrew Talcott, late U. S. Engineers,	19 Jan.	1838

1080. Charles G. B. Daubeny, M.D., of Oxford, England,   19 Jan.   1838
1081. Henry Reed, of Pennsylvania,   19 Jan.   1838
1082. William Norris, of Pennsylvania,   19 Jan.   1838
1084. William Harris, M.D., of Pennsylvania,   20 April 1838
1085. Robert Treat Paine, of Massachusetts,   20 April 1838
1087. *Hugh S. Legaré, of South Carolina,   20 April 1838
1088. Samuel Breck, of Pennsylvania,   20 April 1838
1089. Col. Sylvanus Thayer, U. S. Engineers,   20 April 1838
1090. Francis Wayland, D.D., of Rhode Island,   20 April 1838
1091. Henry Baldwin, of Pennsylvania,   20 April 1838
1092. William H. Prescott, of Massachusetts,   20 April 1838
1094. John Edwards Holbrook, M.D., of South Carolina,   18 Jan.   1839
1095. John C. Cresson, of Pennsylvania,   18 Jan.   1839
1096. James C. Booth, of Pennsylvania,   18 Jan.   1839
1097. Edward Coles, of Pennsylvania,   18 Jan.   1839
1098. J. F. Encke, of Berlin,   18 Jan.   1839
1099. A. Quetelet, of Brussels,   18 Jan.   1839
1100. Humphrey Lloyd, D.D., of Dublin,   19 April 1839
1101. James K. Paulding, of New York,   19 April 1839
1102. John Ludlow, D.D., of Pennsylvania,   19 April 1839
1103. Benjamin W. Richards, of Pennsylvania,   19 April 1839
1104. George W. Bethune, D.D., of Pennsylvania,   19 April 1839
1105. George M. Justice, of Pennsylvania,   19 April 1839
1106. T. Romeyn Beck, M.D., of New York,   19 July   1839
1107. Richard C. Taylor, of Pennsylvania,   19 July   1839
1108. Thomas U. Walter, of Pennsylvania,   18 Oct.   1839
1109. John Penington, of Pennsylvania,   18 Oct.   1839
1111. Charles Rttmker, of Hamburg,   18 Oct.   1839
1112. Capt. John Washington, R. N.   18 Oct.   1839
1113. Rev. Charles Gutzlaff, of Macao,   18 Oct.   1839
1114. Prof. Elias Loomis, of Ohio,   18 Oct.   1839
1115. Prof. Stephen Alexander, of New Jersey,   18 Oct.   1839
1116. Judah Dobson, of Pennsylvania,   17 Jan.   1840
1117. John Forbes, M.D., of England,   17 Jan.   1840
1118. Michael Faraday, of London,   17 Jan.   1840
1119. C. R. Demme, D.D., of Pennsylvania,   17 Jan.   1840

---

* Died June 1843

1120. John J. Vanderkemp, of Pennsylvania,	17 Jan.	1840
1121. Rev. Philip Milledoler, of New Jersey,	17 Jan.	1840
1122. Pedro de Angelis, of Buenos Ayres,	17 Jan.	1840
1123. Isaac Wayne, of Pennsylvania,	17 Jan.	1840
1124. Samuel D. Ingham, of Pennsylvania,	17 Jan.	1840
1125. George M. Dallas, of Pennsylvania,	17 Jan.	1840
1126. Martin H. Boye, of Pennsylvania,	17 Jan.	1840
1127. Hartman Kuhn, of Pennsylvania,	17 April	1840
1128. F. W. Bessel, of Prussia,	17 April	1840
1130. Rev. William H. Furness, of Pennsylvania,	17 April	1840
1131. Capt. Francis Beaufort, of London,	17 April	1840
1132. Paul Beck Goddard, M.D., of Pennsylvania,	17 April	1840
1133. Prof. William H. C. Bartlett, of the U. S. Mil. Acad.	17 April	1840
1134. George M. Wharton, of Pennsylvania,	17 April	1840
1135. George Washington Smith, of Pennsylvania,	17 April	1840
1136. Robert Were Fox, of England,	17 July	1840
1137. John Sanderson, of Pennsylvania,	17 July	1840
1138. Francisco Martinez de la Rosa, of Spain,	17 July	1840
1139. Major James D. Graham, U. S. Topog. Engineers,	17 July	1840
1140. J. B. B. Eyries, of Paris;	17 July	1840
1142. Francois P. G. Guizot; of France,	16 Oct.	1840
1143. Chev. Bernardo Quaranta, of Naples,	15 Jan.	1841
1144. David Irvin, of Wisconsin,	15 Jan.	1841
1145. Adolph C. P. Callisen, of Denmark,	15 Jan.	1841
1146. William Rawle, of Pennsylvania,	15 Jan.	1841
1147. Benjamin Dorr, D.D., of Pennsylvania,	15 Jan.	1841
1148. John L. Stephens, of New York,	15 Jan.	1841
1149. Tobias Wagner, of Pennsylvania,	15 Jan.	1841
1150. Lieut. Col. Edward Sabine, British Army,	16 April	1841
1152. Rev. Roswell Park, of Pennsylvania,	16 April	1841
1153. Robert Christison, M.D., of Edinburgh,	16 April	1841
1154. Prof. Edward Hitchcock, of Massachusetts,	16 April	1841
1155. William Peter, of England,	16 April	1841
1157. George Bancroft, of Massachusetts,	16 July	1841
1158. Alexis de Tocqueville, of Paris,	21 Jan.	1842
1159. Baron de Roenne, of Prussia,	21 Jan.	1842
1160. John F. Frazer, of Pennsylvania,	21 Jan.	1842
1161. E. Otis Kendall, of Pennsylvania,	21 Jan.	1842

1162. Charles Lyell, of London,	21 Jan.	1842
1163. *J. N. Nicollet, of the District of Columbia,	21 Jan.	1842
1164. Baron de Ladoucette, of France,	21 Jan.	1842
1165. E. W. Brayley, of London,	21 Jan.	1842
1166. Stephen Endlicher, of Vienna,	15 April	1842
1167. D. Humphreys Storer, M.D., of Massachusetts,	15 April	1842
1168. Simeon Borden, of Massachusetts,	15 April	1842
1169. Petty Vaughan, of London,	15 July	1842
1170. Frederick Fraley, of Pennsylvania,	15 July	1842
1171. Rev. George Peacock, of England,	21 Oct.	1842
1172. J. I. Clark Hare, of Pennsylvania,	21 Oct.	1842
1173. Prof. Benjamin Peirce, of Massachusetts,	21 Oct.	1842
1174. Leopold II. Grand Duke of Tuscany,	20 Jan.	1843
1175. Louis Agassiz, of Neufchatel,	20 Jan.	1843
1176. William W. Gerhard, M.D., of Pennsylvania,	20 Jan.	1843
1177. Lieut. Col. William Reid, of Bermuda,	20 Jan.	1843
1178. Thomas P. Cope, of Pennsylvania,	20 Jan.	1843
1179. John Lenthall, of Pennsylvania,	20 Jan.	1843
1180. Solomon W. Roberts, of Pennsylvania,	20 Jan.	1843
1181. Ellwood Morris, of Pennsylvania,	20 Jan.	1843
1182. Charles Ellett, of Pennsylvania,	20 Jan.	1843
1183. Charles B. Trego, of Pennsylvania,	20 Jan.	1843
1184. Chevalier Mustoxidi, of Corfu,	20 Jan.	1843
1185. Capt. Charles Wilkes, U. S. Navy,	21 April	1843
1186. Charles M'Euen, of Pennsylvania,	21 April	1843

---

* The intelligence of Mr. Nicollet's death reaches Philadelphia as these pages are issuing from the press.

INDEX

259